Praise for
Essential Prefab Straw Bale Building

The next generation of straw bale buildings is coming, and now we have a guide
for builders, designers, and would-be manufacturers — from one of the best sources
anywhere. As always, Chris Magwood brings enthusiasm, experience and solid science
to the subject, so buy this book! Now! And prepare to learn, smile,
and be part of a cool new generation of building!

— Bruce King, Ecological Building Network, author, *Design of Straw Bale Building*

Essential Prefab Straw Bale Construction is a jumping off point both for professionals
pondering unique, turn-key business opportunities and for novice homeowner-builders
seeking an easier DIY approach to straw bale building. Accelerate your own
research and development by several years in an easy weekend read.

— Ben Polley, Evolve Builders Group

This essential guide covers it all, from the why and how to the nitty gritty details.
Having explored a variety of modular and prefab delivery methods, with the help
of this book, we foresee prefab straw bale becoming a favorite go-to solution."

— Anni Tilt, AIA & David Arkin, AIA, LEED AP / Arkin Tilt Architects

Not satisfied with the impact he's already made, Chris Magwood has
given us a blueprint for building regenerative housing for the masses. No more
baby steps — this book is a giant leap forward for straw bale building!

— David Lanfear, Bale on Bale Construction

[Magwood is] way ahead of the curve as usual with this publication.
And yet it is not a minute too early for the public to be introduced to this important
advancement in the world of low carbon construction. Enclosed you will find
everything you want to know about prefabricated straw bale construction
and a bunch of other things you should know as well.

— Ben Graham, New Frameworks Natural Building

Chris Magwood has been on the forefront of natural building and he once again shows his ability to bring things to the next level. Prefab straw bale may indeed be the catalyst that this incredible technology needs and Chris lays out the plan in detail in this book. From why it's important to how to get started, Chris inspires us to think bigger than ourselves. He takes us deeper into the actual "how-to" details in a way that allows the reader to step forward with confidence and bring the dream of prefab straw bale into reality.

— Andrew Morrison, www.StrawBale.com

While it is not possible to create a standardized, code-approved individual straw bale, the use of prefabricated straw bale panels manufactured in a quality-controlled environment offers a system that will have predictable design parameters. As a researcher, educator and practicing engineer I believe that Magwood has provided us with a practical reference that can provide guidance for students and design professionals who know that the future of sustainable building will be found in creative alternative building strategies.

— Kris J. Dick, Ph.D, P.Eng., Associate Professor and Director,
The Alternative Village, Department of Biosystems Engineering,
University of Manitoba

A book that should be in the hands of every builder, designer, architect, code official and home owner who wants to see better buildings built. Incredibly accessible and technically thorough, it is not often that a technical guide is also such an inspirational read. Every builder, designer, architect, building official and environmentalist will benefit from reading and re-reading this book.

— Melinda Zytaruk, Fourth Pig Green and Natural Construction

Essential Prefab Straw Bale Building just blew the doors wide open for straw bale builders. Building on his success with his book *Making Better Buildings,* Chris Magwood has again written a game-changing book for us natural building geeks. Straw bale construction has been criticized as a niche-y, one-off kind of housing typology. Panels are an incredible solution to scalability and replicability of building with straw. My gratitude goes to Chris again for doing the research and testing for us!

— Emily Niehaus, founder/director, Community Rebuilds

essential
PREFAB STRAW BALE CONSTRUCTION

sustainable
building
essentials

essential
PREFAB STRAW BALE CONSTRUCTION
the complete **step-by-step** guide

Chris Magwood

new society
PUBLISHERS

New Society
Sustainable Building Essentials Series

Series editors
Chris Magwood and Jen Feigin

Title list

Essential Hempcrete Construction, Chris Magwood

Essential Prefab Straw Bale Construction, Chris Magwood

Essential Building Science, Jacob Deva Racusin

See www.newsociety.com/SBES for a complete list of new and forthcoming series titles.

THE SUSTAINABLE BUILDING ESSENTIALS SERIES covers the full range of natural and green building techniques with a focus on sustainable materials and methods and code compliance. Firmly rooted in sound building science and drawing on decades of experience, these large-format, highly illustrated manuals deliver comprehensive, practical guidance from leading experts using a well-organized step-by-step approach. Whether your interest is foundations, walls, insulation, mechanical systems or final finishes, these unique books present the essential information on each topic including:

- Material specifications, testing and building code references
- Plan drawings for all common applications
- Tool lists and complete installation instructions
- Finishing, maintenance and renovation techniques
- Budgeting and labor estimates
- Additional resources

Written by the world's leading sustainable builders, designers and engineers, these succinct, user-friendly handbooks are indispensable tools for any project where accurate and reliable information is key to success. GET THE ESSENTIALS!

Cover design by Diane McIntosh.

Cover art provided by the author. Straw image: AdobeStock_64243755. Thumbs up art: AdobeStock_23490949

Printed in Canada. First printing June 2016.

Funded by the Government of Canada Financé par le gouvernement du Canada | **Canadä**

Paperback ISBN: 978-0-86571-820-3
eISBN: 978-1-55092-614-9
Series ISBN: 978-0-86571-821-0

Inquiries regarding requests to reprint all or part of *Essential Prefab Straw Bale Building* should be addressed to New Society Publishers at the address below. To order directly from the publishers, please call toll-free (North America) 1-800-567-6772, or order online at www.newsociety.com

Any other inquiries can be directed by mail to:
New Society Publishers
P.O. Box 189, Gabriola Island, BC V0R 1X0, Canada
(250) 247-9737

New Society Publishers' mission is to publish books that contribute in fundamental ways to building an ecologically sustainable and just society, and to do so with the least possible impact on the environment, in a manner that models this vision. We are committed to doing this not just through education, but through action. The interior pages of our bound books are printed on Forest Stewardship Council®-registered acid-free paper that is **100% post-consumer recycled** (100% old growth forest-free), processed chlorine-free, and printed with vegetable-based, low-VOC inks, with covers produced using FSC®-registered stock. New Society also works to reduce its carbon footprint, and purchases carbon offsets based on an annual audit to ensure a carbon neutral footprint. For further information, or to browse our full list of books and purchase securely, visit our website at: www.newsociety.com

Library and Archives Canada Cataloguing in Publication

Magwood, Chris, author
Essential prefabricated straw bale construction : the complete step-by-step guide / Chris Magwood.

(Sustainable building essentials)
Includes bibliographical references and index.
Issued in print and electronic formats.
ISBN 978-0-86571-820-3 (paperback).--ISBN 978-1-55092-614-9 (ebook)

1. Straw bale houses--Design and construction--Handbooks, manuals, etc. 2. Walls--Design and construction--Handbooks, manuals, etc. I. Title. II. Title: Prefabricated straw bale construction. III. Series: Sustainable building essentials

TH4818.S77M27 2016   693'.997   C2016-903508-5
                                              C2016-903509-3

Contents

Acknowledgments

I AM EXTREMELY LUCKY AND GRATEFUL that the "rock stars" of the natural building world are simultaneously true superstars of forward thinking about the care of people and the planet *and* humble, approachable, brilliant and most often hilarious people who have built an extended family that I value greatly.

I am equally blessed to have a supportive and caring and patient and engaged immediate family. Together, they have made me who I am, and there aren't really thanks enough for that.

⁓⁓⁓

This person loves
being in charge of the
workingsof a community.
This one loves
the ways that heated iron can be
shaped with a hammer.
Each has been given
a strong desire for certain work.
A love for those motions,
and all motion is love.

— *Rumi*

Chapter 1

Introduction

STRAW BALE BUILDING has made remarkable inroads since the resurgence of this American pioneer building method in the 1990s. From just a handful of enthusiastic builders in the southwestern United States, the use of straw bale walls to make healthy, energy-efficient and low-impact homes and commercial buildings has grown exponentially in only 20 years, and is now recognized in *Appendix S* of the 2015 *International Residential Code (IRC)* in the United States. Professional builders, guilds and organizations around the world are actively building and developing methodologies for meeting code requirements while honoring the environmental impetus that inspired the straw bale pioneers.

A Grassroots Development

In a building industry in which innovation and code change are predominantly driven by patents and corporate financial interests, the fact that straw bale building has grown so quickly and widely within a code-driven context speaks volumes about the inherent applicability of this style of construction to a wide range of climates and building contexts. Those pushing straw bale building forward have done so with only grassroots means at hand; there have been no well-connected lobbyists or industry reps involved, no reliable sources of funding, only individual designers and builders with a passion for the advantages this building style can offer individuals and society at large.

The labor equation

The major hurdle that straw bale building faces in its move toward widespread adoption is the higher labor inputs that site-built straw bale walls require. On a construction-cost basis, it has been adequately demonstrated that a well-trained straw bale building crew can complete homes at a price point that is entirely competitive with conventional approaches, as lower material costs can be leveraged to offset higher labor costs. But the conventional construction industry prioritizes the reduction of on-site labor inputs, even when the cost advantages of labor-saving materials and systems are negligible — or even more expensive. The ability to shorten the build cycle and minimize the time construction crews need to be on site motivates the industry. And if the additional labor time involved in straw bale building and the attendant increase in the length of time that crews are on site are not deterrent enough for mainstream contractors, then the need to train crews to work with a new material, using new techniques and methodologies, are certainly barriers to bale walls becoming part of the mainstream palette of options.

An innovation is born, accidentally

Many straw bale builders have recognized this "deficiency" and have put their minds to exploring ways to get past this hurdle. My own foray into prefabricating straw bale walls began in 2000, when I was invited to create a display for the Toronto Home Show. Realizing it would be impossible to site-build a model straw bale home within the window of time allowed by the show, my partners and I decided to build and plaster the walls in my barn, and deliver the finished panels to the show floor for assembly.

We had no idea if it would work, but went ahead and gave it a try. The panels arrived at their destination in perfect condition — no plaster cracks or other transportation issues, and one panel even survived a forklift mishap that saw it fall to the floor but sustain almost no damage. The demo building was assembled very quickly, and I can still feel the "high" that came along with watching this crazy idea prove itself to be extremely feasible.

In the following 15 years, I've been fortunate to have had the opportunity to build a large number of buildings using prefabricated straw bale panels. Each outing has involved changes to the construction, the lifting and assembling process and the final finishing of the panels. While I cannot say that the process is perfected, I can definitely say that every option we've used has been viable and a vast improvement on the labor input and construction timeline of site-built straw bale walls.

An international phenomenon

At the same time that I began experimenting with prefab bale walls, the idea surfaced in no fewer than three countries on three different continents, without any direct contact between those trying out the system. Within a few years, there would be early adopters in at least half a dozen more countries. Today, there are several commercial ventures producing panels and a number of builders who use the technique on a regular basis.

A long list of advantages

The benefits of prefabricating straw bale walls are many, and include significant reduction of labor input (especially plastering labor); removal of weather restraints during construction; better curing conditions for plaster; more consistent wall construction and plaster thickness; greatly

The first set of prefab bale panels were used to build a small demonstration home in Toronto in 2000.

shortened installation time; and a straighter, squarer and more conventional aesthetic. All of these help to position this type of straw bale building as a real alternative for conventional contractors. A contractor can order walls and have them installed without needing to engage with new supply chains or teach crews new skills. There are no truckloads of bales to be brought to construction sites and stored, no messy on-site plastering. The walls arrive and go up, and within a day or two the wall phase of the building is done and the crew can move on to roof framing, with the structure, insulation, air sealing and sometimes even the final finish of the walls completed in a single step.

Prefab doesn't preclude site building

Some straw bale builders see the use of panelized systems as being antithetical to the grassroots nature of straw bale construction. But there is no inherent either/or choice to be made. Those who have a passion for building with bales on site and hand plastering the walls *in situ* have many excellent reasons to continue to do so. Panelized systems don't preclude site building; they just extend the option to use low-impact, highly insulating bale walls to a range of builders and owners who otherwise would not consider the idea. And panelized systems can also be helpful to owner-builders who do not have access to a large labor pool: panels can be built on site by one or two people and tipped up in place, providing an option for those who can only build part time by breaking the process down into manageable sections that are protected from the weather as soon as they are stood up in place.

Small-scale production facilities

To me, one of the most exciting aspects of prefabricating straw bale walls is that the process favors the use of regional "micro-factories" rather than large, centralized production facilities. The investment required to start a panel-building facility is remarkably small — only requiring basic tools and equipment — and locally acquired straw bales, lumber and plaster materials are the most cost-effective options. Short transportation distances between production facility and building site also make economic and environmental sense. Prefabricated straw bale wall micro-factories are viable local business opportunities in communities anywhere in the world where there is a source of straw bales. Rural communities could benefit from the value-added production and supply of walls to nearby urban centers.

Practicality and idealism combined

As a designer and a builder, I'm a pragmatist as well as a strident environmentalist. Prefabricated straw bale wall panels — or S-SIPs (Strawbale Structural Insulated Panels), as they will be called in this book — are among a very small number of approaches to building that satisfy

Any small shop space can make enough panels for multiple houses, using only basic tools and equipment.

both of these perspectives. In marrying all of the real and measurable ecological advantages inherent in straw bale building with the very practical need to make buildings that are affordable and simple to construct, this approach is one that is ready and waiting to help redirect the building industry in a much more environmentally responsible direction. Whether you are an owner-builder looking to build a single home or a developer looking for a way to green your next subdivision, I hope that this book inspires you in a meaningful way. There is ample evidence that straw bale walls work and work very well; now there is a way to bring those advantages to the modern construction marketplace.

Prefab panels are an exciting way to bring a promising building system to a mainstream practice.
CREDIT: DAN EARLE

Chapter 2
Rationale

Why Straw Bales?

WHILE IT MAY NOT INITIALLY STRIKE one as possessing ideal qualities for construction, there are many compelling reasons for using straw bales as a building material.

Great thermal performance

Though some enthusiasts seem to imbue straw with almost-magical insulating qualities, the truth is that typical straw bales do not have remarkable thermal properties. Most testing has shown R-values ranging from 1.5–2.3 per inch,[1] lower than many commercially produced insulation products. However, straw bale *walls* are widely attributed a whole-wall R-value of 30, which exceeds all current requirements in the International Residential Code (US) and the National Building Code (Canada). A more thorough examination of R-values is presented in the Material Specifications chapter.

However, these numbers don't tell the whole story because bale walls have real-world performance even better than their nominal R-value would indicate. There are several reasons for this, including a near-perfect distribution of thermal mass on the interior and exterior face of the wall, minimal thermal bridges in the wall, and a naturally airtight barrier on both sides of the wall. (See Building Science Notes chapter.) These factors combine to give excellent performance in a wide range of climates.

Good, cheap fill

Straw happens to be relatively inexpensive compared to manufactured insulation products and it comes in relatively large ready-made bundles.

Table 2.1

Material	Approximate cost* for 1 square foot @ ~R-30	Thickness of insulation
Straw bales	$0.75 @ $3.00/bale $1.45 @ $6.25/bale	14–16 inches
Mineral wool batt	$1.40	7 inches
Fiberglass batt	$1.20	9.5 inches
Denim batt	$1.80	8 inches
Dense packed cellulose	$1.45	9 inches
Extruded polystyrene foam	$4.40	5.75 inches
Expanded polystyrene foam	$4.20	7.5 inches
	* cost are averages from building supply retailer Websites, 2015	

The wide straw bale wall (anywhere from 14–26 inches wide for standard bale sizes) insulates very well for a low cost. A square foot (0.09 m²) of bale wall at an approximate R-value of 30 would cost between $0.75–1.45. Bulk purchasing directly from farms can provide lower costs.

Decent structural qualities

Unlike the batt insulation materials in the cost chart shown here, straw bales have a density that allows them to play a structural role in the building. The straw bales in a wall are not the primary structural element of the wall; that role is handled by the vastly stiffer plaster skins[2] for some types of S-SIP. Structural sheathing and/or structural frames can give other types of panel the required structural rigidity.

Bales can be stacked without the need for any framing and still keep their form, and they can act as an excellent substrate for plaster, eliminating a number of components compared to conventional insulation types. They can even hold up a roof temporarily, and (with the right design) absorb earthquake forces for a surprisingly long time — giving a bale wall a resilience that can't be found in other insulation materials.

A by-product produced in vast quantities

Grain farming produces tens of millions of tons of straw annually. Each year, enough straw is produced in North America to build hundreds of thousands of homes. In the U.S., 54–56 million acres of wheat are planted annually,[3] which could produce 6–7 billion construction-grade straw bales, or about 15 million homes each year. And this is only a single grain crop; there are many other viable crops for making straw bales.

Good carbon sequestration

Approximately 40–50% of the mass of a straw bale is carbon.[4] At the code minimum of 6.5 pounds per cubic foot ($100kg/m^3$), every 14"×18"×32" (355×457×800 mm) straw bale contains about 12–15 pounds (5.44–6.8 kg) of carbon. So, a typical 4×8-foot straw panel at the minimum bale density contains about 82–102 pounds (37–46 kg) of carbon. This is carbon

that has been taken out of the atmosphere in a single growing season, and it will be contained in the wall of the building for a significant period of time. Wooden elements used in the construction of the panel also sequester carbon at a similar ratio of mass to weight.

As the harvesting and manufacturing of straw bales has a carbon footprint that is tiny compared to other insulation materials, this volume of carbon tied up in a straw bale panel can add up to a significant reduction of atmospheric carbon. A single 2,000 square foot ($186 m^2$) home would typically use 40 straw bale panels, which means it would sequester approximately 3280–4080 pounds (1488–1850 kg) of carbon!

In Canada, about 200,000 new homes are built each year. If they were all built with straw bale panels, an astonishing 328,000–408,000 tons (298–370 million kg) of carbon could be sequestered annually. That would be a sizable contribution to meeting greenhouse gas reduction targets.

Nontoxic building material

Unlike most other building insulation materials, straw is very benign. Harvesting and baling do not involve any industrial processes or

Table 2.2

Material	Embodied carbon by weight*	Embodied carbon for 4x8 foot wall @ R-28**	Carbon footprint after sequestration
Straw bales	0.063 kgCO$_2$e/kg[5]	8 kgCO$_2$e	-42.8 kg per panel
Mineral wool batt	1.28 kgCO$_2$e/kg	21.75 kgCO$_2$e	21.75 kg per panel
Fiberglass batt	1.35 kgCO$_2$e/kg	17.6 kgCO$_2$e	17.6 kg per panel
Denim batt	1.5 kgCO$_2$e/kg	22.45 kgCO$_2$e	15.45 kg per panel
Dense packed cellulose	0.63 kgCO$_2$e/kg	41.3 kgCO$_2$e	-10.3 kg per panel
Extruded polystyrene foam	3.42 kgCO$_2$e/kg	38.5 kgCO$_2$e	38.5 kg per panel
Expanded polystyrene foam	3.29 kgCO$_2$e/kg	37.25 kgCO$_2$e	37.25 kg per panel
	* figures from Inventory of Carbon and Energy (ICE) 2.0	**material densities from *Making Better Buildings*	

chemicals, and nothing is added to the straw in the bale. The straw does not off gas any toxins and remains a stable, inert natural material in the walls of the building. However, bales can contain dust and particulate, especially if stored in a barn, and appropriate breathing protection should be used during installation.

Why in Panelized Form?

Most of the benefits outlined above are as true of site-built straw bale walls as they are of prefabricated straw bale panels. So why panelize this wall system?

Labor reduction

Relatively high labor input has always been an issue with site-built straw bale walls. The stacking of the bales is not particular labor intensive, but the plastering requires many hours, a lot of scaffolding, and much skill to apply three coats of plaster to all wall surfaces. Wet-process *panels,* on the other hand, can be plastered with the wall in a horizontal position, using the panel framing to create a "container" for the plaster and a screeding surface to level the plaster. Plus, the plaster can be applied in a single coat because there are no issues with slumping and cracking as would happen with a single thick coat applied on a vertical surface. Plastering labor can be reduced by as much as 75% for panels as compared to site-built walls.

Site management

Straw bales are a bulky material, and coordinating their delivery to job sites — especially urban sites — is difficult. On-site storage of bales is highly inconvenient and messy, as large amounts of loose straw accumulate and spread around the job site. On-site plastering is also a very messy process, with large amounts of dropped plaster requiring thorough masking of

all surfaces in and around the building and a lot of cleanup labor. By arriving on site with the wall system already finished, these hurdles are overcome, which could even promote the adoption of straw bale walls in more densely populated areas.

Predictability

Poor weather conditions don't slow down or halt a project, because the walls can be built indoors. Rain and cold temperatures don't affect the quality of the walls, and no on-site time is spent tarping or heating a project to try to keep a project moving.

Shortened build cycle

Panelized walls are installed very quickly — much faster than the on-site construction of any wall system. This greatly reduces the length of the on-site build cycle, helping projects to be completed faster and lowering costs for the builder.

No retraining or reskilling of on-site builders

A panelized straw bale wall does not require any special knowledge of straw bale construction or plastering for installation, allowing crews trained in general residential or commercial construction to take responsibility for t he installation process, and presenting them with walls that accept doors, windows and roof framing in conventional ways.

Who Would Want to Build with Panels?

There are many potential markets for prefabricated straw bale wall panels. Regardless of the market being addressed, one of the most attractive elements of this approach is the low cost of getting started. The initial investment in tools and machinery is very low, and only a flat and (preferably) covered space is needed. Getting started in the business of building S-SIPs has an impressively low entry threshold.

Stand-alone manufacturers

Several companies around the world have formed as straw bale wall panel manufacturers, specializing only in the building and installation of this type of wall system. With energy efficiency, carbon footprint, environmental impact and the high cost of labor being key issues in the construction industry, opportunities exist for companies with a product that can address all of these issues in a cost-effective manner.

Construction companies

Builders in the residential and commercial markets can benefit from building their own prefabricated wall panels. Crew members can build walls during slow seasons or during downtime, keeping workers productive. A stock of ready-to-install walls can allow more jobs to be completed, especially in climates with a limited building season or unpredictable precipitation patterns.

Owner-builders

A panelized system can allow an owner-builder to create a straw bale wall system without the need for a large crew. Wall panels can be built one at a time on the building foundation and tipped up into place at any pace or on any schedule.

Farmers

Wall panels can be built on the farm where the straw is grown, giving farmers a value-added option for their straw bales. Panels can be built during slow seasons or bad weather to maintain productivity and income diversity.

First Nations

Many First Nations are experiencing housing shortages. Panelized straw bale walls can be built in the community they are intended to serve, reducing construction costs and providing employment. Wall manufacturing could continue to be a source of revenue once immediate demand has been satisfied.

Opportunities for Innovation

It is early days for prefabricated straw bale wall panels. Most markets are wide-open to be served, and the technology and processes for building the panels are in their infancy. Current panel builders are using a lot of tools and materials borrowed from other types of construction, and none are employing a high degree of automation. All are working with field-produced bales, but the development of an on-site bale press that converts low-cost bulk straw (from jumbo round or square bales) into precisely sized, consistent bales would provide the significant advantages of consistency and further labor reduction. This is an exciting field for those with a passion for creativity and refinement of processes.

Notes:

1. *Design of Straw Bale Buildings.* Bruce King, Green Building Press, 2006.
2. Ibid.
3. United States Dept. of Agriculture, Economic Research Service, Wheat Data Yearbook 2016.
4. "Carbon sequestration in European soils through straw incorporation: Limitations and alternatives." D.S. Powlson et al.
5. "Establishing a Methodology for Carbon Sequestration in Cotton Production in the US," Lanier Nalley et al.

Chapter 3
Material Properties and Appropriate Use

General Use Parameters

PREFABRICATED STRAW BALE WALLS can be used as exterior and/or interior walls in most low-rise (three story or less) construction scenarios described in the International Residential Code (US) or in Part 9 of the National Building Code (Canada).

The *2015 International Residential Code* includes *Appendix S — Strawbale Construction,* which outlines prescriptions and performance for site-built straw bale walls that can be applied to prefabricated straw bale panels in most cases. The inclusion of this Appendix in the U.S.-based codes could be of great help to anyone producing prefab bale panels by simplifying the process of obtaining building permits.

Suitability for larger projects and/or projects covered by other codes must be determined by the appropriate design professionals. To date, the panels have been used successfully as curtain walls in larger projects, up to six stories in height.

The panels can generally be considered to have structural characteristics that are equivalent to those of conventional wood frame wall systems and can be used in scenarios where wood frame walls are considered by local codes to provide sufficient structural integrity. Structural analysis for particular applications must be considered (see Testing Data in Resources). Panels can be used as structural load-bearing walls or in conjunction with a structural frame as infill walls.

Load-bearing walls

Prefabricated straw bale walls can typically be used as load-bearing exterior and/or interior

Prefabrication makes large straw bale buildings feasible, like the Gateway Building at the University of Nottingham by MAKE Architects.

walls in any scenario in which conventional frame walls are used. In these scenarios, the walls carry all the dead and live loads and transfer these loads to the foundation.

Infill walls

Prefabricated straw bale walls can be used as non-load bearing exterior and/or interior walls in any scenario in which a structural frame carries building loads and in which curtain walls are required. Infill walls can include those in which the frame of the prefabricated panel is designed to carry all structural loads without relying on the straw bale/plaster element.

Specific exclusions

Prefabricated straw bale wall panels should never be used as exterior walls in any below-grade applications, whether used as load bearing or in-fill. Use in flood plain areas should be restricted to elevations above anticipated 100-year flood levels.

Traditional timber frames can be wrapped with S-SIPs in a very straightforward marriage of techniques.
CREDIT: CHRIS MAGWOOD

Load bearing S-SIPs will receive all roof loads.
CREDIT: DAN EARLE

Chapter 4

Building Science Notes

The principles of building science can and should be applied to all elements of a building enclosure throughout the design and construction phases. Managing the flow of heat, air and moisture across the wall system is the key to creating durable, healthy, high performance walls. Viewed through this lens, the wall system is composed of four different *control layers*:

- Thermal control
- Air control
- Vapor control
- Water control

Thermal Control: Principle

The movement of heat is always from an area of higher concentration to an area of lower concentration (from hot to cold). The *thermal control layer* slows the movement of heat through the wall in order to preserve a comfortable interior temperature.

There are three different modes of heat transfer: conduction, convection and radiation, and an effective thermal control layer must work to adequately reduce the amount of all three types of transfer. For a given climate, the thermal control layer must meet a standard of thermal control required by building codes or voluntary energy efficiency standards.

The effectiveness of this thermal control layer is most often measured by the static-state conductivity (U-value) or resistance (R-value). However, testing for thermal resistance at steady-state temperatures and in a moderate temperature range does not always reflect real-world dynamic performance.

Any breaks in the thermal control layer can cause degradation of performance. These "thermal bridges" are caused by materials within the wall assembly (most typically framing members) that have a poorer thermal resistance than the insulation, providing a "short-circuit" for the flow of heat by conduction through the wall.

Thermal control: Application for S-SIPs

1. ***Radiation heat energy emits from any source of heat, including the sun, heating devices, human bodies, or other bodies of warm mass.***

   - The materials struck by this radiation absorb the heat energy. For straw bale walls, the plaster skin or sheathing will be struck by radiant heat waves and will absorb heat energy until its temperature matches that of the radiant source. Plaster, drywall and MgO board are dense materials and will absorb greater quantities of heat before reaching equilibrium than insulated sheathing. Regardless of the mass, the sheathing material will pass this heat energy to the insulation by conduction and convection.

2. ***Conductive heat energy moves by direct contact between materials.***

   - Plaster/sheathing touches the straw bale and the individual pieces of straw make random contact with one another across the thickness of the wall. Each point of intersection facilitates the movement of heat. The low density of straw makes it a relatively poor conductor. Steady state testing gives results of R1.5 to R2.3 per inch (R21–35 for 14-inch (355 mm) and R24–40 for 16-inch (400 mm) bales).

   - The use of insulated sheathing on panels will increase the overall conductive resistance performance of the wall.

   - The straw bale insulation in a panel is typically uninterrupted by framing or other thermal bridges. The frame can be designed to have the same or similar conductive resistance as the bales; if not, they will be thermal bridges that lower the overall thermal performance of the wall assembly.

3. ***Convective heat movement occurs via the movement of air around and inside the wall.***

   - Air on either side of the wall rises as it is warmed and falls as it is cooled. These air currents make direct contact with the wall surface, imparting heat to or removing heat from the wall by conduction. The more air movement, the greater the amount of heat transfer.

4. ***Within the wall itself, convective currents can occur, as the plaster/sheathing and straw surfaces impart heat to the small pockets of air that exist, creating small convective cycles that speed up the transfer of heat.***

   - Straw bales have more open air space than manufactured insulation products, resulting in poorer thermal performance per thickness of material due to convection. For this reason, panel builders should choose very densely packed bales and ensure that all gaps between bales are tightly stuffed with straw or other insulation.

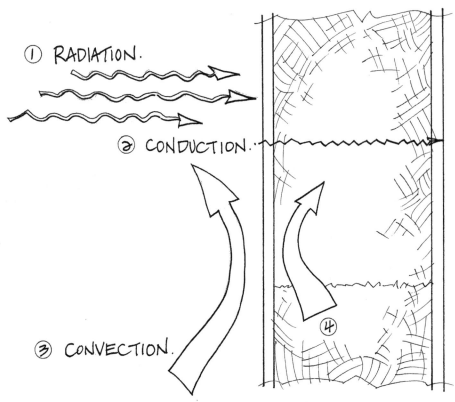

① RADIATION.

② CONDUCTION.

③ CONVECTION.

④

Air control and thermal performance: Principle

The movement of air through a wall assembly "short-circuits" the effect of the insulation, as heat is carried through the wall at an accelerated rate, drastically lowering the thermal performance, regardless of the R-value of the insulation in the wall. Even at a relatively low pressure difference between inside and outside of 10Pa, heat flow through the wall can be nearly 25–50% higher than the R-value of the wall would be with no air movement.

An effective *air control layer* (or layers) must work as a continuous barrier to the free movement of air. Each element of the building enclosure must have an air control layer, and the seams between elements must also be designed and built to prevent air leakage. The continuity of the air control layer is a crucial detail that must be considered at the design stage and integrated into the construction.

Continuity of the air control layer must also be maintained at every penetration through a wall or other element, including electrical outlets, pipe and service conduits.

Table 4.1[1]

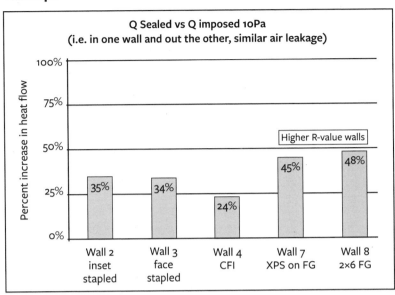

When there is a pressure difference on either side of a wall, heat flow through the wall can be dramatic. Losses can be very high unless there is control of these leaks, with effective R-values decreasing by as much as 48%.

Air control and thermal performance: Application for S-SIPs

Prefabricated straw bale panels have an inherently robust air control layer on both the inside and outside of the wall in the form of the plaster and/or structural sheathing that spans continuously across the face of the insulation and makes contact with the frame around the perimeter of the panel. These materials are thick, durable air control materials. In order not to lose the air control advantages the materials provide, the point of contact between the frame and the plaster/sheathing is an important junction to be air sealed.

Panels can be built with the plaster/sheathing meeting the face of the frame, in which case, a seal can be made using adhesive or caulking. Panels can also be built with the plaster/sheathing meeting the inside edge of the frame. This can be more difficult to air seal reliably, especially with plaster that will shrink away from the frame as it dries/cures. Caulking or expanding tape at this seam can do the job if carefully applied.

The many seams around edges of the panel are important to consider in the design phase and to detail properly during installation.

At the junction with the foundation and the roof and for each panel-to-panel seam, a system for preventing air migration must be designed and implemented. The details of these junctions will vary depending on the style of frame and the materials they are meeting, but the walls will not meet their intended level of thermal performance if each of these seams is not appropriately sealed to provide a continuous air barrier.

Air control and moisture performance: Principle

The movement of air through a wall assembly can carry a significant amount of water vapor into the wall. Warmer air holds a higher quantity of water vapor and will be driven toward areas of colder temperatures. As this air cools down, its ability to hold water vapor decreases, and it will at some point deposit its moisture load in the wall. This can cause serious issues of mold, rot and deterioration of the wall materials over time.

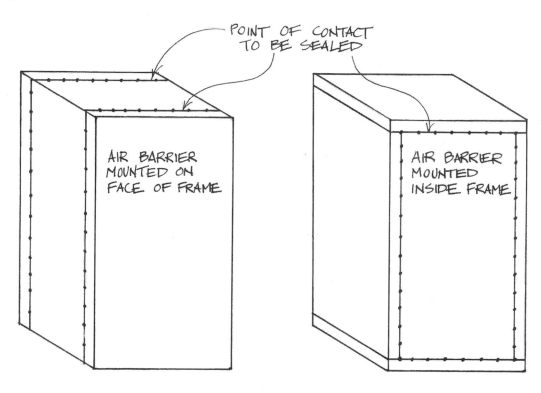

POINT OF CONTACT TO BE SEALED

AIR BARRIER MOUNTED ON FACE OF FRAME

AIR BARRIER MOUNTED INSIDE FRAME

Relatively small holes through which air can enter a wall assembly can carry large amounts of moisture into the wall. Even a hole as small as 1×1 inch (25×25 mm) can carry as much as 7.5 gallons (28 liters) of moisture into a wall over the length of a heating season (figures for Ottawa, Canada), making air sealing a vital part of controlling moisture migration issues.

Air control and moisture performance: Application for S-SIPs

Controlling moisture that is moving into the wall via air leakage should be handled in the same manner as described for Air Control and Thermal Performance. If all potential leakage points in each panel and at all panel seams are addressed and air leakage is controlled, the moisture being carried into the wall will be reduced to levels that are not of concern in a straw bale wall system.

Vapor control: Principle

In many ways, the air control layer in a wall system performs the function of a *vapor control layer*. The vast majority of water entering a wall system is carried by free-moving air (see above). If air movement is controlled, a small amount of moisture will still enter the wall by diffusion.

In the diffusion process, vapor pressure caused by a difference in moisture content on either side of a wall causes moisture to migrate at a molecular level through pore spaces in the wall materials. The ability of a material to resist this diffusion of moisture is measured by units called "perms." There are three classes of vapor retarders, based on perm ratings:

Class I — 0.1 perm or less (qualifies as a vapor barrier, or vapor *impermeable*)

Class II — 0.1 to 1.0 perms (vapor *semi-impermeable*)

Class III — 1.0 to 10 perms (vapor *semi-permeable*, or vapor *permeable* — 10 perms or greater)[2]

Class IV — 10 perms or greater (vapor *permeable*)

Materials with a perm value higher than 10 perms would be considered vapor *permeable*.

Straw bale wall systems do not incorporate vapor barriers, and are best built using Class III or fully vapor permeable materials in the assembly.

Vapor control: Application for S-SIPs

A straw bale wall system is built using assemblies of materials that have a very high degree of moisture storage capacity. Straw is a *hygroscopic* material, meaning it has a large volume of pore spaces able to *ad*sorb vapor on all the surface area within the material. For a bale with a density of 8 pounds per cubic foot (128 kg/m^3), more than 1 pound or 0.12 gallons (0.45 kg or 0.45 liters) of water in vapor form can safely be stored per square foot (0.09 m^2) of wall area.[3] This is a much higher storage capacity than required for the moisture deposited into the wall via diffusion (0.0026 gallons per square foot), and even capable of handling moisture loading from gross air leakage. At 0.9 gallons per 1×1 inch hole, a straw bale can theoretically store the water deposited if there were a hole that size in every square foot of wall. Various plaster and sheathing options for straw bale panels are also able to safely store a large volume of adsorbed water. Maintaining the moisture balance in a wall system is the key to long-term performance and durability.

Since we know that the materials in a straw bale panel have a high safe storage capacity, as long as a reasonable drying potential exists, then the moisture balance will remain in the safe working zone.

Drying of straw bale panels will occur largely via diffusion, as vapor pressure draws moisture through the wall from to the side with the lowest relative humidity (from inside to outside when it's cold outside, from out to in, if it's warmer outside). If the materials in the panel are sufficiently permeable to allow this drying to occur, then a straw bale panel system will be feasible. An 18-inch (450 mm) straw bale has a perm rating of 3–6 perms.[4] The balance of the permeability will depend on the perm rating of the plaster and/or sheathing materials. Ranges of permeability for plaster are summarized in the table shown here.

Any plaster type with a perm rating higher than 4 will be suitable for panels. Dry sheathing materials should likewise have perm ratings of 4 or higher. See manufacturer ratings to ensure suitability for vapor control.

Table 4.2: Permeance of plaster skins

	US perms	Metric perm
Typical vapor barrier (by definition)	<1	<60
1:3 cement:sand (1.5")	1	50
5:1:15 cement:lime:sand (1.5")	4	200
1:1:6 cement:lime:sand (1.5")	7	400
1:2:9 cement:lime:sand (1.5")	9	500
1:3 lime:sand (2")	9	500
Earth plaster (2")	11	600

Water control: Principle

Moisture accumulation from rain is the highest risk for walls. The design of a building should consider this risk and mitigate exposure as much as is practically possible. Roof overhangs, foundation toe-ups and proper window flashing and sills are all key elements of water control.

Inevitably, the wall will receive some degree of exposure to water on the exterior. The degree of protection required will depend upon climate conditions (amount and frequency of rain, strength and direction of winds, drying conditions). Protection can come in the form of *surface protection* and/or *rain-screen protection*.

Surface protection comes from the final cladding material applied to the exterior face of the wall, or a combination of the cladding material and a protective coating (typically paint).

A *rain screen* is made by creating a ventilated space in front of the main air/weather barrier on the exterior face of the wall, and affixing a final cladding material in front of this ventilated space. This allows the vast majority of precipitation to be shed by the cladding, with the ventilation space allowing the cladding and the wall assembly a drying channel that is protected from the elements. This is a more resilient, durable approach than surface protection, but it adds another layer of material to the building, increasing costs.

Water control: Application for S-SIPs

Protection from precipitation loading for straw bale panels can come two forms:

Surface protection of plaster — Where plaster applied directly to the straw bales will be the finished exterior surface, it is possible for the plaster to become super saturated by precipitation, resulting in moisture loading of the straw behind the plaster. In regions with less precipitation and/or good drying conditions, the surface protection offered by the plaster may be adequate.

Where the climate includes a lot of precipitation and/or poor drying conditions, the plaster itself may not offer enough protection against precipitation. In these conditions, a surface treatment on the plaster is required. The best option is a water-repellant paint with a Class III or high perm rating (10 perms or more), such as potassium silicate paint. Using a protective coating with a Class I or II perm rating will cause vapor control moisture issues and should never be used.

Rain screen — A rain screen can be added to the face of a panelized straw bale wall system, built over a plaster substrate or a structural sheathing material.

Vertical strapping applied to the face of the wall creates open channels that keep moisture-laden cladding from touching the wall and provide a pathway for drying air to carry away moisture from the cladding and the wall. These channels must be protected from intrusion by insects or animals.

Notes:

1. Source: /greenbuildingadvisor.com/blogs/ dept/musings/air-leakage-degrades-thermal- performance-walls

2. Vapor retarder classes per U.S. Dept. of Energy's Building America research and included the International Building Code, International Residential Code and International Energy Conservation Code

3. "Building Science for Strawbale Buildings." John Straube, *Building Science Digest,* 112, 2009.

4. Ibid.

Chapter 5

Material Specifications

Straw Bales

A T THE HEART OF ANY PREFABRICATED straw bale wall panel are the bales themselves. The bales are composed of the stems of cereal grains, and are harvested after the plant has fully matured and the valuable seed-heads have been removed. The remaining straw is taken from the field and pressed into rectangular bales that are tied with strings or wires.

All straw can be broadly categorized as *lignocellulose*, or plant dry matter. Each variety of straw has different concentrations of cellulose, hemicellulose and lignin. Most varieties of straw have been used for building, including wheat, rice, oat, barley, rye, spelt, hemp and switchgrass.

It may be helpful to think of each stalk of straw as a tiny tree; the chemical composition, strength and characteristics are very similar to that of softwood species. This comparison can be particularly useful when considering the durability of straw, as the lifespan of straw is the same as that of wood and it is prone to deterioration under the same conditions as wood or other cellulose-based building materials.

Specifications

There are no binding standards for building bales. Straw bales used in prefabricated walls should meet the minimum requirements of AS103, *Appendix S — Strawbale Construction, 2015 International Residential Code.*

Permeability

A straw bale has a level of permeance of 2 to 4 US perms.[1] By these figures, straw bale

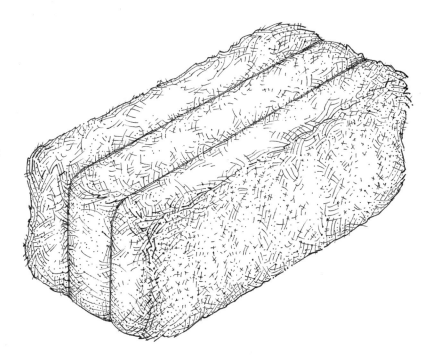

insulation can be considered a Class III vapor retarder, though it is more often considered a vapor permeable material. A straw bale wall with bales of a density of 8 pounds per cubic foot (128 kg/m^3) can safely store 1 pound (0.45 kg) of water in vapor form per square foot (0.09 m^2) of wall area.[2] This storage capacity gives the straw bale insulation the functional properties of a fully vapor permeable material.

Thermal performance

Thermal tests conducted in labs worldwide have shown R-values ranging from 1.5–2.8 per inch of thickness for a straw bale wall.[3] The figures at both the high and low end of this range have largely been seen as anomalies due to poor testing methods. A figure of R-2.1–2.3 per inch seems to be consistent across thorough testing results. *Appendix S* of the *2015 IRC* accepts an R-value of 30 (2.14 per inch) for a typical straw bale wall. A testing program in Denmark found similar results of R-31.5 (2.25 per inch) for a two-string-bale wall.[4] Testing in the UK attributes straw bale wall assemblies a value of R-31.85 (2.28 per inch).[5]

Energy modeling of straw bale buildings at the design phase has shown that calculations made using R-30 as the assumed thermal performance value for S-SIP walls results in real energy use figures that corroborate this figure as being accurate.

Sourcing

There aren't any established sourcing pathways for acquiring bales in the construction industry. Direct contact with a farmer producing any type of grain can secure a supply of bales. However, if the production of panels is intended to be for multiple homes every season, it may be best to contact a bale wholesaler that carries a large inventory of bales year round. While it is NOT hay that you want to build with, hay bale wholesalers tend to also deal in straw bales.

Farm supply stores and cooperatives are a good place to make connections with farmers who have a quantity of straw bales for sale. There are also many regional online listings for farmers selling bales and straw wholesalers.

Cost

The price of straw varies widely by region, type of straw, size and density of the bale and size of the harvest. Small square bales of straw currently retail for $2.50–7.00 per bale, with the average price around $4.00 per bale. Placed on-edge in an S-SIP, a single bale will represent 3.5–4.5 square feet (0.33–0.42 m^2) per bale. At the average price, bales will cost $0.88–1.15 per square foot ($9.47–12.37 per m^2) but can range from a low of $0.71 to as much as $2.00 per square foot ($7.64–21.50 per m^2). As the price of straw is so variable, this cost should be well researched early in the process of designing S-SIPs.

Straw can often be pre-purchased prior to the harvest by contacting farmers in the spring and agreeing on a price and quantity of bales in advance. This can help to remove some uncertainty regarding the price and availability of straw.

Table 5.1: Environmental impacts of straw bales

Ecosystem Impacts	Embodied Energy	Carbon Footprint	Indoor Environment	Waste
Low to Moderate. Impacts largely the result of monoculture agriculture, including fertilizer, herbicide and pesticide use. Confirm practices with straw bale source to verify degree of impacts. Straw from organically grown crops will have the lowest impacts.	**Very low.** 0.24 MJ/kg* or 3.5–4.0 MJ per average two-string bale. No high heat processes required. Production energy input split with embodied energy of cereal grain production.	**Very low.** 0.06 kgCO₂e/kg* or 0.2 kgCO₂e/kg per average two-string bale. Production carbon output split with carbon output of cereal grain production. High carbon sequestration potential.	**Very low to Low.** Very low surface toxicity. No toxic off gassing. Material separated from interior air by plaster or sheathing.	**Very low to Low.** Construction: Leftover or unused straw is fully compostable. Polypropylene strings may be recycled in some jurisdictions. End of life: Straw is fully compostable. Embedded mesh will require separation.

Note:
* Data is from Inventory of Carbon and Energy (ICE) 2.0, University of Bath.

Plaster Skins

Plaster skins used in prefabricated walls should meet the minimum requirements of *Appendix S — Strawbale Construction, 2015 International Residential Code*. Where applicable, IRC sections R702 and/or R703 and National Building Code (NBC) of Canada sections 9.28.5 and 9.28.6 may apply. However, these code specifications are designed to cover the mixture and application of plaster to substrates with properties that are significantly different from straw bales, and many provisions about surface preparation will not be directly applicable.

Clay Plaster

Clay or "earthen" plaster is any coating that relies on clay as the primary binder. Clay plasters are not recognized in the IRC (US) or the NBC (Canada), but are specified in *Appendix S, Strawbale Construction, 2015 International Residential Code.*

Composition

Recipes for clay plasters can vary widely, particularly those made from naturally occurring clay soils. In general, these plasters are a mix of clay binder (pure clay in dry bagged form, or a high clay content soil), a well distributed range of aggregate (ranging from silt [0.00016 inch or 0.004 mm] to large sand [3/16 inch or 5 mm] particle sizes) and a high percentage of fiber (typically chopped straw, but can also be other natural fibers including sisal and hemp, or animal hair, in a range of sizes to a maximum of 1½ inch [40 mm]).

Thickness

Clay plasters should have a minimum thickness of 1 inch (25 mm).

Strength

Clay plasters should have a minimum compressive strength of 100 psi (0.69MPa) when completely dry as per *IRC Appendix S*, Table AS106.6.1.

Permeability

~11 US perms (600 metric perms).[6]

Sourcing

Clay The clay content of the plaster can come from dry, bagged clays that are typically sourced from pottery supply outlets, though mortar clay can be obtained from some masonry supply outlets. Bulk orders are best placed directly with the manufacturer, as pottery supply outlets typically sell in small quantities at higher costs.

Bulk supply of clay can also come from quarries that specialize in clay mixes. These are often used for sporting fields and southern climate road bases. Such suppliers will be able to give you accurate information regarding clay and aggregate content, and deliver in large quantities. Some suppliers are capable of producing custom mixes to suit your specifications. In southern regions, road base may contain a suitable amount of clay and be available in bulk.

Clay soils can be used for earthen plasters. An ideal soil would have 20–40% clay content, with the remainder of the soil being a wide distribution of aggregate, from silt (less than 5%) up to small pebbles (3/8 inch or 9.5 mm). Soils with a less than ideal blend of clay and aggregate can be

---

BASIC FORMULATION (BY VOLUME)

1 part clay or high-clay content soil

1–2 parts sharp aggregate

0.5–1 part fiber

Water to provide a plastic, workable mix

amended with additional clay and/or aggregate to achieve the minimum working strength of 100 psi (0.69 MPa).

Aggregate The aggregate component of clay plaster can come from the typical construction aggregate supply chain. The ideal aggregate is similar to "concrete sand" or "stucco sand," with grading performed according to ASTM C33.

Table 5.2: Fine-aggregate Grading Limits
(ASTM C 33/AASHTO M 6)

Sieve size		Percent passing by mass	
9.5 mm	(3/8 in.)	100	
4.75 mm	(No. 4)	95 to 100	
2.36 mm	(No. 8)	80 to 100	
1.18 mm	(No. 16)	50 to 85	
600 μm	(No. 30)	25 to 60	
300 μm	(No. 50)	5 to 30	(AASHTO 10 to 30)
150 μm	(No. 100)	0 to 10	(AASHTO 2 to 10)

Fiber The fiber content of a clay plaster is important to the mix, and straw bale builders have been tending toward mixes with very high fiber content, often equaling one third or more of the total mix by volume. In this type of mix, the fiber is not just an additive to lend some tensile strength to the mix; it becomes a large percentage of the overall aggregate content. A mix with a high percentage of fiber allows for a greater concentration of clay in the plaster without the excessive cracking that would appear in a clay-rich mix that relied only on aggregate. Higher clay content can raise the working strength of the plaster and help it to resist erosion from rain exposure and prevent the dusting that can occur if too much aggregate is used.

The simplest source for natural fiber is chopped straw, since the same bales used in the walls can be used to make fiber, as can all the leftover straw from the production of the panels. The straw can be run through a hammer mill or chipper/shredder with a ¾–1 inch (19–25 mm) screen. Other natural fibers, such as hemp, sisal and wool can be purchased in bulk, but will be costlier than straw.

Cost

Bagged dry clay currently retails for $14–25 for a 50-pound (22.5 kg) bag. At a typical clay plaster thickness of 1.5 inches (38 mm), the cost would be $0.86–1.56 per square foot of wall area ($9.26–16.80 per m²). Sand will need to be added to bagged clay to make a plaster, at an average cost of $0.09–0.12 per square foot ($0.97–1.29 per m²).

Bulk sports field clay currently retails for $35–75 per ton. These mixes can be ordered with the ideal amount of clay and aggregate together, and at a thickness of 1.5 inches (38 mm) would cost $0.32–0.66 per square foot of wall area ($3.44–7.10 per m²). If additional sand is required, there will be a slight cost increase.

Clay soils can often be obtained for free, or at a very low cost. Contact local excavators to find out about pricing and availability. If bagged clay or sand is needed to amend the clay soil, this will be an additional cost.

Fiber obtained from chopped straw will cost approximately $0.05–0.10 per square foot ($0.54–1.08 per m²). The volume of chopped straw can vary widely, and must be determined before a definitive cost can be set.

These cost estimates are based on S-SIPs built with clay plaster on both sides of the wall. For panels with clay plaster on only one side, the costs would, of course, be halved.

Table 5.3: Environmental impacts of clay plaster

Ecosystem Impacts	Embodied Energy	Carbon Footprint	Indoor Environment	Waste
Very low to Low. Site soils can be excavated and used with minimal disturbance to site. Processed or bagged clay impacts will vary with extraction and refinement processes. Confirm practices with source to verify degree of impacts. Large-scale quarrying can cause habitat destruction and surface and ground water interference and contamination.	**Very low.** 0.083 MJ/kg* or 0.58 MJ/m² (for excavating and transporting soil locally). No figures available for bagged clay products, but figures will be higher due to energy used for crushing and drying clay.	**Very low.** 0.0052 kgCO₂e/kg* or 0.108 kgCO₂e/m² (for excavating and transporting soil locally). No figures available for bagged clay products, but figures will be higher due to energy used for crushing and drying clay.	**Low to Moderate.** Clay plasters typically contribute to high indoor air quality. Contaminants in soil or bagged clay may be difficult to assess and pose some difficulty in confirming air quality impact. Hygroscopic qualities of clay provide excellent moisture handling characteristics.	**Very low.** Construction: Leftover plaster can be returned to the ground with no impact. End of life: No ill impacts. Embedded mesh will require separation.

Note:
* Data is from Inventory of Carbon and Energy (ICE) 2.0, University of Bath.

Lime and Lime-cement Plaster

Lime plaster includes formulas that use either hydraulic or hydrated lime as the binder.

Lime-cement plaster includes formulas that use a blend of hydrated lime with Portland cement as the binder, with no more than 50% of the binder being Portland cement.

In general, these plasters use a ratio of binder to well-graded aggregate in the range of 1 part binder to 2–4 parts aggregate. Fiber is a common additive in these mixes when applied to straw bales, and can include poly fibers, chopped straw or animal hair.

> **BASIC LIME PLASTER FORMULATION (BY VOLUME)**
>
> 1 part hydrated lime or hydraulic lime (NHL 3.5 or 5)
>
> 2–3 parts sharp aggregate
>
> 0–1 part fiber
>
> Water to provide a plastic, workable mix

Thickness

Lime plaster should have a maximum thickness of ½ inch (16 mm) per coat.

Strength

Lime plaster should have a compressive strength of 600 psi (4.14 MPa) at 28 days as per *IRC Appendix S*, Table AS106.6.1.

> **BASIC LIME-CEMENT PLASTER FORMULATION (BY VOLUME)**
>
> 1 part hydrated lime
>
> 0.25–1 part Portland cement (S-, N- or O-type mixes)
>
> 4–6 parts sharp aggregate
>
> 0–1 part fiber
>
> Water to provide a plastic, workable mix

Thickness

Lime-cement plaster should have a maximum thickness of 1 inch (25 mm) per coat.

Strength

Lime-cement plaster should have a compressive strength of 1,000 psi (6.9 MPa) at 28 days as per *IRC Appendix S*, Table AS106.6.1.

Permeability

1:1:6 cement:lime:sand — 7 US perms (400 metric perms)

1:3 lime:sand — 9 US perms (500 metric perms)[7]

Sourcing Lime

There are two types of lime for plasters: *hydrated* and *hydraulic*.

Hydrated lime cures chemically by re-absorbing the carbon dioxide that was driven out of the limestone during production. In order for this process to happen properly, the lime must have contact with the atmosphere in order to have access to the CO_2; typically, lime plasters are applied in successive thin coats to allow each coat a reasonable amount of time to carbonize. Hydrated lime does not work for wet-process panels where the full thickness of the plaster is applied in a single, thick coat. Hydrated lime is widely available through masonry supply outlets in North America. Type S lime is the most common product to use for lime plaster.

Hydraulic lime cures chemically as well, partly by reacting with water to create a new compound, and partly by carbonizing. The hydraulic set of this type of lime means it can be used for the thicker coats common to wet-process panels, though it is unlikely that the working strength achieved by successive thinner coats will be reached. A natural hydraulic lime (NHL) with a hydraulic reaction of 5 should be used for thick applications in prefab panels.

Hydraulic lime is not produced in North America. There are distribution networks established for imported European natural hydraulic lime. In larger cities, masonry supply outlets may stock hydraulic lime, or it may require special ordering through smaller masonry supply centers.

Cement Portland cement is widely available and regularly stocked at building and masonry supply outlets.

Cost

Hydrated lime prices range from $10–15 for a 50-pound (22.5 kg) bag. At a thickness of 1 inch on both sides of a panel, the cost will be $0.16–0.25 per square foot of wall ($1.72–2.69 per m²). Sand will need to be added to hydrated lime to make a plaster, at an average cost of $0.09–0.12 per square foot ($0.97–1.29 per m²).

Hydraulic lime price can vary widely depending on sources, with retail listings ranging from $24–46 for a 55-pound bag. At a thickness of 1 inch on both sides of a panel, the cost will be $0.40–0.77 per square foot of wall ($4.30–8.29 per m²). Sand will need to be added to hydraulic lime to make a plaster, at an average cost of $0.09–0.12 per square foot ($0.97–1.29 per m²).

Cement-lime is commonly purchased as a pre-mix product at retail costs of $5.80–8.00 for a 50-pound (22.5 kg) bag. At a thickness of 1 inch on both sides of a panel, cost is $0.10–0.14 per square foot ($1.08–1.51 per m²). Sand will need to be added to hydraulic lime to make a plaster, at an average cost of $0.09–0.12 per square foot ($0.97–1.29 per m²).

Table 5.4: Environmental impacts of of lime and lime-cement

Ecosystem Impacts	Embodied Energy	Carbon Footprint	Indoor Environment	Waste
Moderate. Limestone is a non-renewable resource but is abundantly available. Large-scale quarrying can cause habitat destruction and surface and ground water interference and contamination.	**High.** 1.11 MJ/kg* or 116.9 MJ/m². Lime and/or cement are processed at high temperature, in addition to quarrying and crushing energy input.	**High.** 0.174 kgCO$_2$e/kg* or 18.33 kgCO$_2$e/m². Lime will absorb CO$_2$ during the curing process, but due to fuel use during processing will still be a net carbon emitter, though accurate figures are difficult to assess.	**Very low.** Lime-based plasters can contribute to high indoor air quality, providing naturally antiseptic qualities and no toxic off gassing.	**Low to Moderate.** Construction: Plasters can be left in the environment or crushed to make aggregate. End of life: Plasters can be left in the environment or crushed to make aggregate. Embedded mesh will require separation.

Note:
* Data is from Inventory of Carbon and Energy (ICE) 2.0, University of Bath.

Wood Framing

A wood frame of some sort is a key element in all styles of prefabricated straw bale wall panels. The size and orientation of the wood used in the frame will vary depending on the load path through the wall panel and the degree to which the wood frame is expected to support the loads. Some wood frames are intended to handle all of the structural loads for the walls, and others rely on the bales, plaster and/or sheathing to provide some degree of the structural capacity.

Strength and Dimensions

Wood framing members for S-SIPs should conform to the standards of IRC R602 (US) or NBC 9.3.2 (Canada). For use of wood in a structural role that does not conform to code prescriptions, *The Wood Handbook: Wood as an Engineering Material* from the Forest Products Laboratory of the USDA (FPL-GTE-190) or *The Engineering Guide for Wood Frame Construction* from the Canadian Wood Council or similar guidelines should be used.

Sourcing

Wood framing materials can be sourced from conventional building supply outlets or sawmills. Those with third party sustainability certifications are likely to have lesser ecosystem impacts.

Cost

Wood framing retail costs range from $0.32–0.51 per lineal foot ($1.05–1.67 per m) for 2×4s to $0.50–0.63 per lineal foot ($1.64–2.07 per m) for 2×6s. A typical 4×8 foot S-SIP framed with 2×4 lumber would require 48 lineal feet (14.6 m) of lumber and would cost $15.36–24.48.

The use and arrangement of lumber in an S-SIP panel will vary according to loads, code requirements and panel type, and costing will need to reflect a specified design.

Table 5.5: Environmental impacts of wood framing

Ecosystem Impacts	Embodied Energy	Carbon Footprint	Indoor Environment	Waste
Low to High. Forestry practices can range from third-party verified sustainable harvesting to unregulated clear cutting. Confirm practices with source to verify degree of impact.	Low. 7.4 MJ/kg* (spruce lumber) or 43.66 MJ per 2x4x8. Quantities of lumber used for different prefabricated wall systems will vary widely, and total embodied energy figures must be assessed based on design.	Low. 0.2 kgCO$_2$e/kg*. Quantities of lumber used for different prefabricated wall systems will vary widely, and total carbon footprint must be assessed based on design. High carbon sequestration potential.	Low. Framing lumber in most panel systems is not in direct contact with indoor air, but softwood lumber does not have toxic off gassing or contain any red list chemicals.	Low to High. Construction: Framing lumber can be utilized strategically to minimize waste, but standard lengths can lead to high percentage of off cuts. Wood waste can be recycled or composted. End of life: Can be recycled or composted. Will require separation from assembly.

Note:
* Data is from Inventory of Carbon and Energy (ICE) 2.0, University of Bath.

Structural Composite Lumber

This family of products includes laminated veneer lumber (LVL), parallel strand lumber (PSL), laminated strand lumber (LSL) and oriented strand lumber (OSL). These manufactured wood products use a variety of wood fiber types bound by adhesives to achieve dimensions, working strengths, and stability that are greater than framing lumber.

In some S-SIP systems, composite lumber products are used as framing members, especially where the frame is intended to carry all or most of the structural loads.

Strength

There are a variety of standards for structural composite lumber products. In general, use should conform to applicable code standards, the *Engineered Wood Construction Guide* or other appropriate design standards.

Dimensions

Structural composite lumber can be ordered in a wide variety of dimensions. Some products will have common dimensions, and most can be ordered in custom sizes.

Thermal performance

Structural composite lumber has R-values in the range of R-0.9 to 1.25 per inch. Frame components that create a continuous bridge across the S-SIP assembly will lower the overall thermal performance of the panel. Frames should be designed to minimize or completely eliminate these thermal bridges to maintain a high level of thermal performance for the wall system.

Permeability

These products are not suitable for use as interior or exterior sheathing due to low permeance ratings of less than 1 US perm.

Sourcing

Manufacturers offer a range of standard product dimensions, and will also create custom dimensions by special order.

Cost

This category covers a wide range of products, each with different pricing points. If using structural composite lumber, pricing will need to be obtained for specific design options.

Table 5.6: Environmental impacts of structural composite lumber

Ecosystem Impacts	Embodied Energy	Carbon Footprint	Indoor Environment	Waste
Low to High. Forestry practices can range from third-party verified sustainable harvesting and production to unregulated clear cutting. Confirm practices with source to verify degree of impact. Harvesting and manufacturing of ingredients for glues have high impacts associated with petrochemicals.	Moderate to High. 15 MJ/kg* or 95.25 MJ/m² at ½ inch. Materials typically undergo several heating processes. Glues also use high processing energy.	Moderate. 0.45 kgCO₂e/kg* or 2.86 kg/CO₂e/m². Wood has moderate to high carbon sequestration potential.	Moderate to High. These products are made using a variety of binder formulas, most containing formaldehyde, among other potentially toxic chemicals. Confirm with supplier regarding binder ingredients and off gassing potential.	Low to High. Construction: Manufactured wood products can be custom ordered to appropriate dimensions. End of life: Cannot be recycled or composted due to adhesives. Will require separation from assembly.
Note: * Data is from Inventory of Carbon and Energy (ICE) 2.0, University of Bath.				

Wood Fiber Sheathing

Wood fiberboard sheet materials can be used as insulative and structural exterior and/or interior sheathing for S-SIPs. These sheet materials are made of compressed wood fibers (typically recycled) with a wax-based binder. Wood fiberboard products used for S-SIPs should conform to the standards of ASTM E72 (Standard Test Methods of Conducting Strength Tests of Panels for Building Construction) and/or ASTM C209 (Standard Test Methods for Cellulosic Fiber Insulating Board) or equivalent. Because they can provide both structural and insulative properties, they are a good option for S-SIPs.

Many fiberboard products are rated for exterior applications, but some are not. Be sure to specify exterior grades if that is required.

Strength

The strength requirements for wood fiberboard sheathing will vary depending on the design of the S-SIP panel. The sheathing may be designed to play a large role in the structural performance of the wall, or it may not figure into the structural calculations at all. The ASTM E72 and/or C209 data can be used to ensure that a particular product will meet the strength requirements.

Should an S-SIP not have anchoring points that match the requirements of established performance standards, engineering principles should be used to determine appropriate fastening methods and loads.

Dimensions

North American products typically conform to standard sheet sizes (4×8 and 4×9 feet). European imports are often nonconforming, and sizes should be confirmed prior to ordering. Panels can come in a wide range of thicknesses, from ½ inch to 4 inches (12.7–101.6 mm). Square edges or tongue-and-groove edges may be available.

Table 5.7: Environmental impacts of wood fiber sheathing

Ecosystem Impacts	Embodied Energy	Carbon Footprint	Indoor Environment	Waste
Low to Moderate. Most wood fiber products are made from post-industrial waste streams and do not directly involve the harvesting of timber. Third-party verified practices can be sourced, and practices should be confirmed with source to verify impacts. Most manufacturers use nontoxic binders; this should be confirmed.	**Moderate.** 9.36MJ/kg* or 71.3 MJ/m² at 1-inch. There are both wet and dry processing methods for wood fiber sheathing, and no third party data is currently available to assess this category thoroughly. The additional energy efficiency added by these types of panels helps off set EE compared to plaster or non-insulating sheathing.	**Low.** There are no available figures for this product category. Particle board is rated at 0.86 $kgCO_2e/kg$* and uses similar processing techniques.	**Low to Moderate.** These products are made using a variety of binder materials. Many manufacturers advertise nontoxic binders, and some have third-party verification for low emissions. Sheathing is not in direct contact with indoor air.	**Low to High.** Construction: Sheathing can be utilized strategically to minimize waste, but standard sizes can lead to high percentage of off cuts. Wood waste can be recycled or composted. End of life: Can be recycled or composted. Will require separation from assembly.

Note:
* Data is from Inventory of Carbon and Energy (ICE) 2.0, University of Bath.

Thermal performance

Wood fiberboard products typically have thermal resistance values of R-2.5 to 3.5 per inch. Check with manufacturer for product-pecific ratings.

Permeability

Thickness and density of products will affect perm ratings. Permeance in the range of 18–28 US perms is common, qualifying them as fully vapor permeable. Wood fiberboard products can be used to create a permeable sheathing on either or both sides of an S-SIP.

Sourcing

Wood fiberboard products are currently well developed in European markets, but less so in North America. The North American Fiberboard Association (www.fiberboard.org) represents the industry, and can be used to source local manufacturers and distributors of products.

European companies have limited distribution in North America, typically through specialized green building product outlets. Fiberboard is a common material used in Passive House projects in Europe, and distributors may be located via local Passive House chapters.

Cost

Price for wood fiberboard will vary by thickness. Products with structural properties at 1½ inch (38 mm) thickness range from $1.00 to 2.00 per square foot ($10.76 to 21.52 per m^2).

Magnesium Oxide Sheathing

Magnesium oxide (MgO) sheathing is a relative newcomer to the construction industry. This sheet material is made from magnesium oxide cement cast into thin panels and cured under proper conditions. Most products use a percentage of wood chips and/or perlite in the mix, and the boards are commonly faced with fiberglass matts on both sides. MgO board has a relatively high structural capacity, excellent fire rating, and is resistant to mold and mildew.

Magnesium oxide sheathing products can be used as an exterior and/or interior sheathing for S-SIPs, in dry or mixed processes.

Strength

The strength requirements for MgO sheathing will vary depending on the design of the S-SIP panel. The sheathing may be designed to play a large role in the structural performance of the wall, or it may not figure into the structural calculations at all. There are no specific ASTM standards for MgO board, but structural capacity can be found via results for ASTM E72 (Standard Test Methods of Conducting Strength Tests of Panels for Building Construction).

Many MgO products have not undergone ASTM or equivalent testing for use as a structural sheathing. While these products may be strong enough, it is best to choose products that have passed structural tests if they are to be used for structural purposes.

Should an S-SIP not have anchoring points that match the requirements of established performance standards, engineering principles should be used to determine appropriate fastening methods and loads.

Dimensions

MgO panels come in a variety of standard sizes. Those intended to be used as structural sheathing are commonly 4×8, 4×9 or 4×10 feet (1220×2400, 1220×2750 and 1220×3050 mm) at ½ and ⅝ inch (12.7 and 16 mm) thicknesses. MgO products used as tile backer are commonly 3×5 feet (915×1525 mm).

Thermal performance

MgO panels do not contribute meaningfully to the thermal performance of an S-SIP. The thermal performance of the board is not rated by any manufacturers at this time. The product can contribute to thermal performance if it is used to provide an effective air control layer for the interior and/or exterior of the panel.

Permeability

Sheathing used on the interior face of S-SIPs should be higher than 2 US perms, and the exterior face should rate higher than 4 US perms. Permeability values for MgO board from different manufacturers range widely, from as low as 0.9 to 7.5 US perms. Using a board with an inappropriately low perm rating, especially on the exterior face of an S-SIP can result in moisture damage, so be sure to confirm permeability with the manufacturer.

Sourcing

Most of the world's magnesium board comes from China, though some North American production is beginning to get underway. Magnesium board is available through some conventional building supply outlets and distributors, though often as a special order and not a stock item. It may be beneficial to go directly to manufacturers/importers.

Quality control has been an issue for many magnesium board suppliers. Due diligence should be performed to ensure that consistent quality is available from the chosen supplier.

Cost

MgO board at ½ inch (12.7 mm) thickness range from $0.80 to $1.20 per square foot.

Table 5.8: Environmental impacts of magnesium oxide sheathing

Ecosystem Impacts	Embodied Energy	Carbon Footprint	Indoor Environment	Waste
Moderate to High. Magnesium carbonate is quarried from surface-based pits, mostly in Asia. It is difficult to obtain accurate information about impacts. Large-scale quarrying can cause habitat destruction and surface and ground water interference and contamination.	**Moderate to High.** 6 MJ/kg* or 56.5 MJ/m² at ½". There are no third party figures available. MgO is heated during processing, resulting in relatively high EE. Fiberglass mesh is integrated, as is perlite. All of these high-intensity materials make the industry figure quoted seem too low.	**Low.** There are no available figures for this product category.	**Low to Moderate.** This product category uses stable, nontoxic basic materials and reports to be free of off gassing and toxic chemicals. There are currently no third party verifications, but MgO products are recommended by certified Bau-Biologists.	**High.** Construction: Sheathing can be utilized strategically to minimize waste, but standard sizes can lead to high percentage of off cuts. Composite material cannot be composted or recycled. End of life: Cannot be recycled or composted. Will require separation from assembly.

Note:

* Data is from Inventory of Carbon and Energy (ICE) 2.0, University of Bath.

Gypsum Sheathing

Gypsum sheathing comes in two forms, one for exterior use (featuring a fiberglass and/or waxed paper coating and a wax-treated gypsum core) and one for interior use (conventional "drywall"). For S-SIPs, any use of interior-rated gypsum will limit the ability of the panel to withstand moisture during transportation and installation.

Strength

Gypsum sheathing products used as an exterior and/or interior sheathing for prefabricated straw bale wall panels — whether in wet- or dry-process panels — should conform to the standards of ASTM C1396 (Standard Specification for Gypsum Board), ASTM C1278 (Standard Specification for Fiber-Reinforced Gypsum Panel), or ASTM C1177 (Standard Specification for Glass Mat Gypsum Substrate for Use as Sheathing).

Should an S-SIP not have anchoring points that match the requirements of established performance standards, engineering principles should be used to determine appropriate fastening methods and loads.

Dimensions

Gypsum sheathing products come in a variety of standard sizes. Those intended to be used as structural sheathing are commonly 4×8, 4×9 or 4×10 feet (1220×2400, 1220×2750 and 1220×3050 mm) at ½ and ⅝-inch (12.7 and 16 mm) thicknesses. Thinner products may be used as an interior finishing layer on S-SIPs.

Thermal performance

Gypsum sheathing products do not contribute meaningfully to the thermal performance of an S-SIP. The thermal performance of ½-inch (12.7 mm) board is typically R-0.45. The product can contribute to thermal performance if it is used to provide an effective air control layer for the interior and/or exterior of the panel.

Permeability

Interior and exterior gypsum products have tested permeance ratings of 20 to 25 US perms, making them suitable for use as a permeable sheathing for S-SIPs.

Sourcing

Gypsum panels are widely available through conventional building supply outlets.

Table 5.9: Environmental impacts of gypsum sheathing

Ecosystem Impacts	Embodied Energy	Carbon Footprint	Indoor Environment	Waste
Moderate. Gypsum is a soft rock quarried from surface-based pits. Large-scale quarrying can cause habitat destruction and surface and ground water interference and contamination.	Moderate to High. 6.75 MJ/kg* or 60.75 MJ/m² at ½ inch. Gypsum is processed using a moderate amount of heat. Fiberglass facing is applied, and may not be included in the above figure.	Moderate. 0.39 kgCO₂e/kg* or 3.51 kgCO₂e/m². Does not include production of fiberglass facing.	Moderate. *Exterior product:* Fiberglass particulate is shed during handling. The material may contain some quantity of toxic chemicals, including vinyl acetate monomer, acetaldehyde and formaldehyde. Product would be used only on exterior of wall. *Interior product:* The paper and glue of interior drywall can be a good medium for mold growth in wet conditions.	High. Construction: Sheathing can be utilized strategically to minimize waste, but standard sizes can lead to high percentage of off cuts. Composite material cannot be composted or recycled. End of life: Cannot be recycled or composted. Will require separation from assembly.

Note:

* Data is from Inventory of Carbon and Energy (ICE) 2.0, University of Bath.

Manufactured Wood Sheathing

This category includes *plywood* and *oriented strand board (OSB)*. These materials are commonly used in conjunction with framing lumber to create frame components for S-SIPs, most often to join individual pieces of framing lumber to span the width of the wall.

Strength

Wood sheathing products that are rated for structural exterior use are suitable as framing elements for S-SIPs. There are a variety of standards for structural composite lumber products. In general, use should conform to applicable code standards, the *Engineered Wood Construction Guide* or other appropriate design standards.

Dimensions

Most products in this category are 4×8 feet (1220×2400 mm) at ⁷⁄₁₆, ½, ⅝ or ¾ inch (11, 12.7, 16 and 19 mm) thicknesses.

Thermal performance

Manufactured wood sheathing has R-values in the range of R-0.9 to 1.25 per inch. Frame components that create a continuous bridge across the S-SIP assembly will lower the overall thermal performance of the panel. Frames should be designed to minimize or completely eliminate these thermal bridges to maintain a high level of thermal performance for the wall system.

Permeability

This type of sheathing material does not have the necessary permeability to be used as structural sheathing on the inside or outside faces of an S-SIP, with exterior grade versions of these products rating less than 1.0 US perm.

Sourcing

These sheet materials are available through conventional building supply outlets.

Cost

Retail prices for ½-inch (12.7 mm) OSB range from $0.26 to 0.32 per square foot ($2.80 to 3.44 per m²), and exterior grade sheathing plywood is $0.42 to 0.53 per square foot ($4.52 to 5.70 per m²).

Table 5.10: Environmental impacts of manufactured wood sheathing

Ecosystem Impacts	Embodied Energy	Carbon Footprint	Indoor Environment	Waste
Low to High. Forestry practices can range from third-party verified sustainable harvesting and production to unregulated clear cutting. Confirm practices with source to verify degree of impact. Harvesting and manufacturing of ingredients for glues have high impacts associated with petrochemicals.	**Moderate to High.** 15 MJ/kg* or 95.25 MJ/m² at ½ inch. Materials typically undergo several heating processes. Glues also use high processing energy.	**Moderate.** 0.45 kgCO₂e/kg* or 2.86 kg/CO₂e/m². Wood has moderate to high carbon sequestration potential.	**Moderate to High.** These products are made using a variety of binder materials. Most binders contain formaldehyde, among other potentially toxic chemicals. Confirm with supplier regarding binder ingredients and off gassing potential.	**Low to High.** Construction: Standard sheet sizes can be used effectively, but custom wall sizes can generate a lot of off cuts. End of life: Cannot be recycled or composted due to adhesives. Will require separation from assembly.

Note:
* Data is from Inventory of Carbon and Energy (ICE) 2.0, University of Bath.

Recycled Sheathing (ReWall)

This material is made from compressed recycled packaging, and can be used in a manner similar to manufactured wood sheathing to join individual pieces of framing lumber to span the width of the wall. As a 100% recycled material with no added adhesives or binders, it can be an attractive replacement for other sheet materials.

Strength

The exterior sheathing product from ReWall has been tested to ASTM E72 (Standard Test Methods of Conducting Strength Tests of Panels for Building Construction) with results that make it suitable for structural purposes.

Dimensions

Standard sheet dimensions are 4×8 or 4×9 feet, with thickness of ½ inch (12.7 mm).

Thermal performance

Though this material would not be used as a key part of the thermal control layer for an S-SIP, it has a better R-value than manufactured wood sheathing, at approximately R-2 per inch. Framing components built with this material spanning the width of the wall would have greatly reduced thermal bridging compared to manufactured wood sheathing.

Permeability

The permeance rating for this material of 0.69 US perms prevents it from being used as a sheathing material for the interior or exterior face of S-SIPs.

A nonstructural version of the panel, called Naked Board, is rated at 2.53 US perms, making it feasible for use as an interior sheathing. The un-faced version of this sheathing has a permeability that would make it feasible for use as an interior structural sheathing.

Sourcing

This product must be obtained directly from the manufacturer or from a limited number of distributors.

Cost

Exterior sheathing grade ReWall retails for $0.42 per square foot.

Notes:

1. *Design of Straw Bale Buildings.* Bruce King, Green Building Press, 2006, pg 155.
2. "Building Science for Strawbale Buildings." John Straube, *Building Science Digest,* 112, 2009.
3. *Design of Straw Bale Buildings,* pg 193.
4. "Straw bale houses: Design and material properties." prepared by Jørgen Munch-Andersen, Birte Møller Andersen, and Danish Building and Urban Research.
5. Shea, A. D., Wall, K. and Walker, P., "Evaluation of the thermal performance of an innovative prefabricated natural plant fibre building system." *Building Services Engineering Research and Technology,* 34 (4), 2013. pp. 369–380.
6. *Design of Straw Bale Buildings.* pg 156.
7. Ibid.

Table 5.11: Environmental impacts of recycled sheathing

Ecosystem Impacts	Embodied Energy	Carbon Footprint	Indoor Environment	Waste
Low. The material is made from 100% post consumer recycled content, largely juice boxes and tetra-paks.	**Low to Moderate.** There is no published EE data for this material. Recycled material is shredded and compressed, using low amounts of input energy.	**Low.** There is no published carbon footprint data for this material. Factory inputs are assumed to be low. The high paper content offers some carbon sequestration potential.	**Low.** Food grade material is used, and no VOC or other emissions are associated with the material.	**Low to Moderate.** Construction: Standard sheet sizes can be used effectively. Off cuts can reenter local recycling streams. End of life: Can reenter local recycling streams.

Chapter 6

Design Options for S-SIPs

THERE ARE MANY POSSIBLE WAYS to choose to build a prefabricated straw bale wall panel. This chapter outlines a variety of options that a designer and/or builder of S-SIPs will need to consider before moving forward with a project. Presented here are general types of construction, based on the work of S-SIP builders around the world; the final details used for any of these types of construction will need to be aligned with local code requirements.

Each section of this chapter can be treated like a menu, and the designer/builder can select the options that best suit the particular conditions for the project. There is no single option that is better than any other. By clearly identifying the project's needs in terms of material costs and availability, labor input, structural and thermal requirements, and handling/installation process, there will be an option to suit anything — from a one-off owner-builder creating panels that are built in place and tipped up to a large-scale manufacturer attempting to automate a process and build large quantities of panels inexpensively.

An S-SIP panel is a combination of elements that together provide the complete structural, thermal, air, moisture and water protection requirements demanded of a wall.

1. **Base plate.** *Can provide:*
   - Containment for straw bales.
   - Attachment to foundation.
   - Fixture point for sheathing, strapping, flashing and trim.
   - Potential lifting point.

2. **Side plate.** *Can provide:*
   - Containment for straw bales.
   - Partial or full load-bearing capacity to transfer loads to foundation.
   - Attachment to adjacent panels.

   - Fixture point for sheathing, strapping and trim.

3. **Top plate.** *Can provide:*
   - Containment for straw bales.
   - Partial or full load-bearing capacity to transfer roof loads to panel and/or side plates.
   - Attachment to floor and roof systems.
   - Attachment to adjacent panels.
   - Fixture point for sheathing, strapping and trim.

4. **Straw bales.** *Can provide:*
   - Insulation.
   - Some degree of structural capacity.
   - Vapor diffusion and moisture storage capacity.

5. **Sheathing.** *Can provide:*
   - Structural capacity for roof and racking loads.
   - Fire protection of straw bale insulation.
   - Air sealing of straw bale insulation.
   - Water and moisture protection of straw bale insulation.
   - Attachment point for strapping, trim and finishes.
   - Final finished surface of the wall.

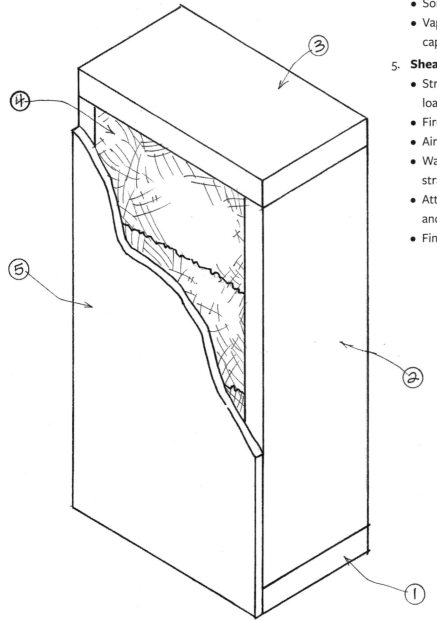

The following sections of this chapter are intended to provide a range of design options for each of these elements.

S–SIP Fabrication Methods

There are essentially three ways to build a pre-fabricated straw bale wall panel: *wet proccess, dry proccess,* and *mixed process.*

Wet process

This type of panel uses wet-applied plaster skins on the straw bale surfaces of the wall, and closely resembles site-built straw bale walls.

Within this type of panel, the structural performance can be based on:

- The model of *load-bearing straw bale walls,* in which the strength of the two stiff plaster skins adhered to the straw bales combines with the strength of the frame to give the wall its structural properties.
- The model of *structurally framed straw bale walls,* in which the frame of the panel is designed to give the wall its structural properties.
- The model of *infill straw bale walls,* in which a separate structural frame is designed to provide a load path for the building, and the panels need only support their own weight.

A coat of plaster is applied to the straw bales in the wet process.
Credit: Katie Howard

Table 6.1

Advantages of Wet Process	Disadvantages of Wet Process
Structural properties can be modeled on existing engineering of straw bale walls	Creates a heavy panel (30–35 lbs/ft² or 145–170 kg/m² typical)
Structural properties of plaster skins can reduce reliance on strength of frame	Plastering adds a complex step to construction process
Excellent sealing and air tightness, less potential for thermal convection loops	Plaster requires multiple components and proper storage conditions
Excellent and well-documented fire rating	Requires curing time before panel can be moved
Excellent and well-documented moisture handling	Joints between plastered panels may require additional site work to make connections
Plaster can be final finish for interior and/or exterior, eliminating additional steps on site	

Dry-process panels have a permeable sheathing material applied over the straw bales. No plaster is used in this type of panel.

Dry process

This type of panel does not use any plaster on the straw bales, but contains the bales within a frame using a permeable, structural sheathing over the bales on the inside and outside face of the wall.

Within this type of panel, the structural performance can be based on:

- The model of *structurally framed straw bale walls,* in which the frame of the panel is designed to give the wall its structural properties. The sheathing material may be able to contribute to the structural performance of the frame.
- The model of *infill straw bale walls,* in which a separate structural frame is designed to provide a load path for the building, and the panels need only support their own weight.

Table 6.2

Advantages of Dry Process	Disadvantages of Dry Process
Lighter panel weight	Permeable sheathing materials can be costly
Shorter assembly time, no curing time	Potential for thermal convection loops between straw and sheathing, may cause loss of thermal performance and moisture issues
Added thermal performance if insulated sheathing is used	Requires stronger, costlier frame
Simpler connections between panels	Sheathing requires additional cost and time for on-site installation of rain-screen siding
Sheathing products more common in conventional construction, may have more market appeal	North American sourcing of permeable structural sheathing can be difficult
Fire rating based on sheathing materials, less fire testing data available for wall system	

Mixed process

This type of panel is a hybrid option that includes a plaster skin applied to the bales (often at a reduced thickness) and a permeable sheathing applied over the bales/plaster. With this type of panel the structural performance is based on the combined strength of the frame, bales/plaster and sheathing.

In this mixed-process panel, plaster is placed onto a permeable sheathing (MgO board shown). The sheathing and plaster bond together, creating a panel that uses plaster to seal the straw but also has an MgO-board final finish.

Table 6.3

Advantages of Mixed Process	Disadvantages of Mixed Process
Structural properties can be modeled on existing engineering of straw bale walls	Creates a heavier panel
Structural properties of plaster skins can reduce reliance on strength of frame	Longer labor time involves plastering and installation of sheathing
Excellent sealing and air tightness, less potential for thermal convection loops	Plaster requires multiple components and proper storage conditions
Excellent and well-documented fire rating	Requires curing time before panel can be moved
Excellent and well-documented moisture handling	Permeable sheathing materials can be costly and difficult to source in North America
Simpler connections between panels	May require stronger, costlier frame
Sheathing products more common in conventional construction, may have more market appeal	Additional costs for both plaster and sheathing materials

Hybrid process

This type of panel includes a plaster skin applied to the bales on one side (often the interior side) and a permeable sheathing applied over the bales/plaster on the other side (typically exterior). With this type of panel the structural performance is based on the combined strength of the frame, bales/plaster and sheathing. The potential for bending must be considered if the two sides of the panel are very different in strength.

Table 6.4

Advantages of Hybrid Process	Disadvantages of Hybrid Process
Hybrid panels allow the builder to incorporate the advantages of a plastered interior wall system, including low cost materials (especially clay plasters), excellent fire protection, the possibility of providing the finished wall surface in one coat, durability and distinct aesthetics. If the plaster is applied to the top face of the panel during construction, this is the less labor intensive side to plaster. The sheathed exterior wall system is less prone to weather damage and able to adapt to a greater range of siding options.	Hybrid panels have the disadvantages associated with both wet- and dry-processes. These panels will require unique analysis of their structural capacity, as the two sides of the panel may have different working strengths and characteristics.

Straw Bales

To date, all manufacturers of prefabricated straw bale wall panels have used bales made by typical farm machinery. The dimensions of these bales can vary between different baling machines, but the rectangular shape and the baling process is common to all bales. Variations in bale use are largely about orientation.

Bales on edge

Bales are laid with their length aligned in the direction of the wall with the strings showing on the interior and exterior face. In general, the straw is aligned vertically in this scenario.

 Advantages:

- Many bales are 14 inches (355 mm) wide, and with 1-inch (25 mm) plaster skins or permeable sheathing, a wall width of 16 inches (406 mm) can be achieved. This is an ideal size for the efficient use of sheet materials as

Bales on edge can be cut to provide custom heights within the panel. CREDIT: DAN EARLE

framing components, as each 4-foot wide sheet will produce three 16-inch strips with no waste.
- Without cutting or modifying bales, this size and orientation makes a wall with the narrowest footprint.
- Bales in this orientation can be cut lengthwise to produce custom wall heights.
- This orientation uses the least number of bales to achieve a particular wall height.
- The strings can be cut on the bales after installation to allow the bales to expand and fill the gaps that can occur where bales meet.

 Disadvantages:

- It can be slightly more difficult to adhere plaster to the string-side faces of the bales if using wet-process.
- Gaps between bales need to be filled from both sides of the wall.

1. Strings visible on inside and outside face of wall.
2. Straw predominantly oriented vertically.
3. Cut and folded ends oriented to top and bottom of wall.
4. Base and side plate of frame.

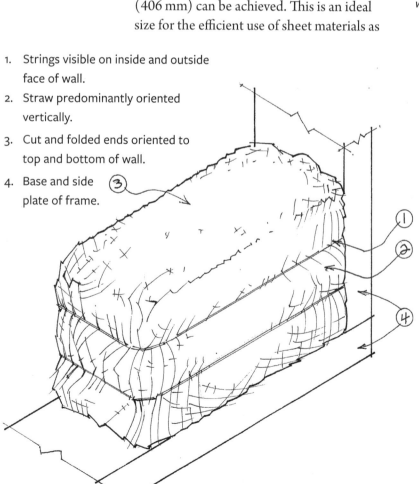

Bales on flat

Bales are laid with their length aligned in the direction of the wall with the strings showing on the top and bottom of the bales. In general, the straw is aligned horizontally in this scenario.

 Advantages:

- Plaster adheres better to this face of the bales for wet-process panels.
- Gaps that occur where bales meet can be easily stuffed across the full width of the wall from one side.

 Disadvantages:

- Bales cannot be altered in height, resulting in walls with fixed height dimensions based on bale size.
- Wall width does not work out evenly with sheet materials, resulting in waste if these materials are used.
- More bales required to achieve a particular wall height.

1. Strings visible on top side of bale.
2. Predominant straw orientation across the width of the wall.
3. Cut and folded sides show to inside and outside face of wall.
4. Base and side plate of frame.

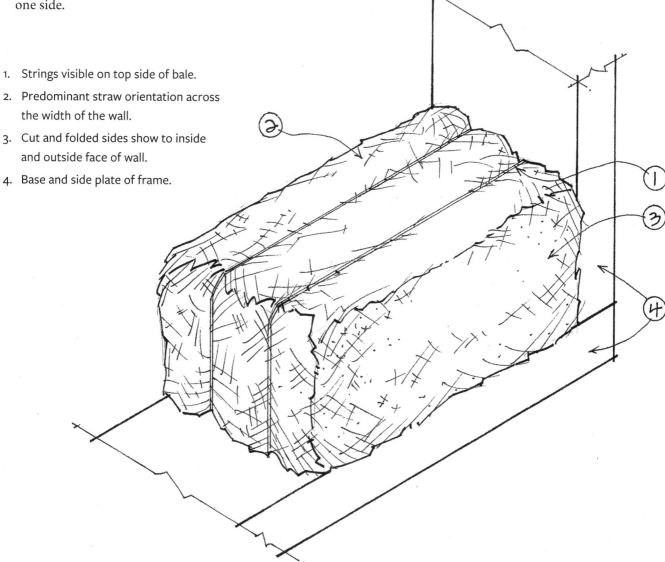

Bales custom cut from jumbo bales

Jumbo bales are commonly produced in a range of sizes, from 30×32×60 inch (760×810×1525 mm) to 48×48×96 inch (1220×1220×2440 mm). These bales can be sliced between the strings to create a number of different sizes of bales.

 Advantages:

- The straw in jumbo bales is much more densely packed, resulting in bales that have better structural and thermal performance.
- The width of the wall can be optimized or varied.
- Straw is often less expensive when purchased in jumbo bales.

 Disadvantages:

- Jumbo bales require machinery to move and manipulate, as they are too heavy to be handled manually.

- The cutting of jumbo bales requires equipment and time.
- The length and width of jumbo bales can be more difficult to adjust due to the higher tension on the strings.

Custom-made bales

For a wall manufacturer producing large numbers of walls, it would be advantageous to invest in a bale press that could make bales to desired dimensions on site, rather than relying on field-made bales. Straw can be purchased in large round or square bales at comparatively low cost, and fed into a customized baling machine that could produce bales with the desired dimensions and density. To date, this has not been done, but baling technology is well developed in other industries and relatively inexpensive; the ability to control the production of bales would offer many advantages to a wall producer.

Frame components

S-SIP frames can be made with a wide range of materials, and these materials can be employed in numerous variations. This section looks at each component that can be used in a frame, and at specific strategies for each type of frame assembly. A prospective panel builder can review all the options outlined here and make decisions about materials and applications based on specific project goals. Many of the established S-SIP manufacturers offer several variations as options.

Base plates

Structural requirements

- Transferring loads to the foundation as needed. This is most important for panels that are intended to be load bearing, with loads transferred evenly along the entire length of the wall.
- Providing attachment points to the foundation to resist all lateral and uplift forces. This will depend on the type of foundation and load paths. It is common to secure panels only at the ends and to fasten to the uprights rather than the base plate.
- Providing attachment points for structural sheathing, if used.
- Handling uplift forces during lifting and placement if the lifting force is applied at the bottom of the wall.

Building science requirements

- Provide a "toe-up" to raise the straw bale insulation above floor level, creating a buffer against flooding inside the building.
- Attach any required flashing and/or ventilation required at the bottom of the wall on the exterior side.
- A moisture barrier and air sealing detail will need to be incorporated into the base plate connection, based on the type of foundation being used.
- Thermal break from interior to exterior. If required to meet performance targets, the base plate may need to incorporate an insulation material or be made from a material with the required thermal resistance.

Functional requirements

- Interior trim attachment surface. The height of the trim may require a certain base plate dimension for secure attachment.
- Possible wiring chase (see page 45). This can be built into the panel, or external wiring chase baseboards may be secured to the base plate.
- Attachment points for decorative finishes on sheathing.

Framing lumber base plates

1. Sheet materials typically ½″ (12.5 mm), can be thicker if required for structural purposes or thinner to save on materials.

2. Top sheet may not be required, bales can sit directly on insulation and framing lumber.

3. Bottom sheet material may be in contact with foundation, use adequate moisture protection and/or moisture resistant sheet material.

4. Framing lumber is typically 2×4 (38×89 mm) or 2×6 (38×140 mm).

5. Framing lumber laid flat is typical.

6. Framing lumber on edge can accommodate bottom-lifting strategies, and may be useful for attaching baseboard trim and/or flashing and siding.

7. Permeable, moisture resistant insulation can include mineral wool, expanded clay or glass balls, perlite, or similar.

8. Cavity should be filled completely.

9. All joints between materials to be caulked or glued to prevent air and moisture leakage.

10. Fastener size and frequency to be determined by structural requirements.

11. Width of base plate to match bale width if plaster or sheathing is intended to cover the frame edge.

12. Width of base plate to match full wall width if plaster or sheathing is to be flush with frame.

13. Notched bottom plate can be used to accommodate lifting straps and/or forklift blades for handling. Gaps can be insulated after installation.

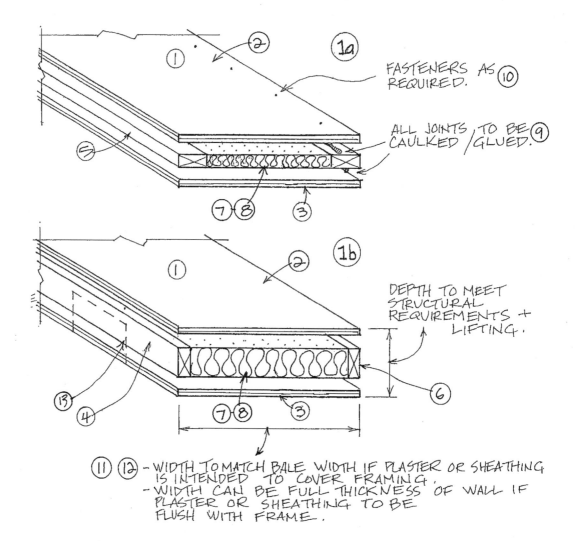

Variation:

The interior framing lumber can be recessed to create a built-in chase for electrical wiring. This can be capped with baseboard trim after installation.

 Advantages:

- Low cost, widely available materials.
- Variations in size and dimension possible with same basic materials.
- Insulated portion prevents or reduces thermal bridging of solid material base plate.
- Structural capacity can be achieved with appropriate design.

 Disadvantages:

- Multiple components to be cut, sized and fastened results in more labor time.
- Mix of materials requires proper attention to sealing, fastening and connecting to prevent air leakage.
- Sheet materials may contain adhesives with formaldehyde other potentially off gassing chemical components.

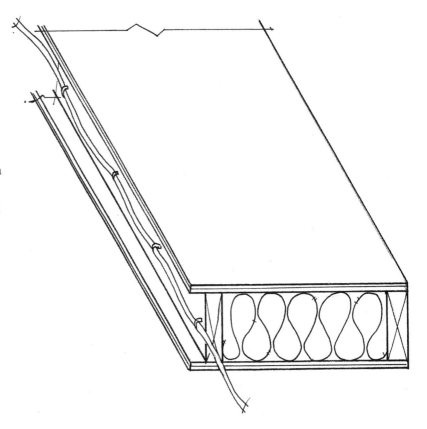

Solid material base plate

1. *Glulam base plate made from solid framing lumber attached with adhesive.*
   - Dimension of lumber dictates sizing.
   - Appropriate moisture protection must be used between glulam plate and foundation.

2. *Cement-bonded wood wool or cement- bonded wood fiber products are available in sheet form and can be cut to size for base plates. These are the only insulated option for solid material frames.*

3. *Laminated veneer lumber (LVL), parallel strand lumber (PSL), laminated strand lumber (LSL) or oriented strand lumber (OSL) made from wood plies or wood strands bonded with adhesive.*
   - Come in a variety of sizes and can be custom ordered to exact dimensions.

- Appropriate moisture protection must be used between plate and foundation.

4. *Width of base plate to match bale width if plaster or sheathing is intended to cover the frame edge.*
   - Width of base plate to match full wall width if plaster or sheathing is to be flush with frame.

👍 *Advantages:*

- Single material minimizes labor input.
- Wood-wool or wood-fiber versions offer good thermal performance with no thermal bridges.

👎 *Disadvantages:*

- Materials must be special ordered and/or custom formed.
- Materials cost more than framing lumber options.
- Solid wood materials create a thermal bridge through the wall.
- Adhesives can contain toxic chemicals and may off gas.

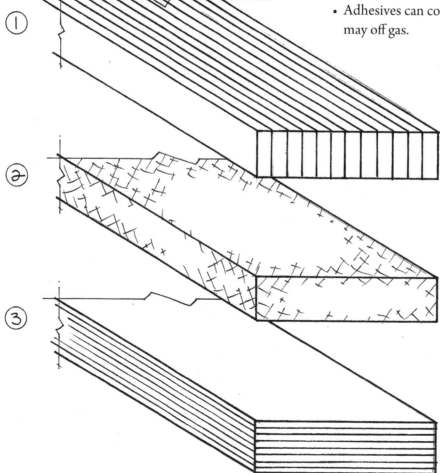

Side posts

Structural requirements

- Transferring loads to the foundation as needed. Some side plate/post options are intended to carry all structural loads to the foundation, and others share load-bearing duties with the bales/plaster/sheathing.
- Providing connection points to the base plates and top plates to resist all uplift forces.
- Providing connection points between panels, if necessary.
- Providing attachment points for structural sheathing, if used.
- Handling uplift forces during lifting and placement if the lifting force is applied at the bottom of the wall.

Building science requirements

- Thermal break from interior to exterior. If required to meet performance targets, the side plate/post may need to incorporate an insulation material or be made from a material with the required thermal resistance.
- Provide adequate surface for air sealing between panels, as required.

Functional requirements

- Providing attachment points for doors, windows and trim, as required.
- Possible wiring chase. This can be built into the panel or external wiring chase baseboards may be secured to the base plate.
- Attachment points for decorative finishes on sheathing.

Framing lumber side posts

 Advantages:

- Low cost, widely available materials.
- Structural capacity can be achieved with appropriate design.
- Variations in size and dimension possible with same basic materials.
- Space between posts can be insulated to prevent thermal bridging.

 Disadvantages:

- Space between posts can be difficult to insulate. Notching bales can be labor intensive and inaccurate. Rigid insulation may be difficult to fasten/secure.
- Unconnected posts offer comparatively low out-of-plane load resistance.

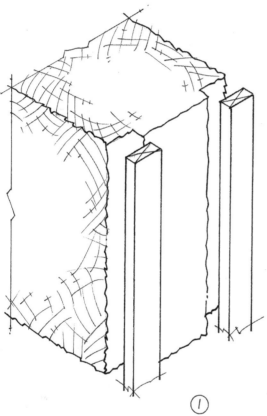

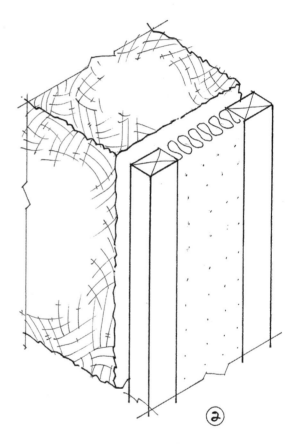

- 2x4 or 2x6 posts spaced at bale width (if plaster/sheathing will cover framing) or wider than bales (if plaster/sheathing is to be contained by posts).
- Bales can be notched around framing (1), or rigid insulation can be cut and inserted to fill void (2).

- 2x4 (or larger) posts spaced at bale width (if plaster/sheathing will cover framing) or wider than bales (if plaster/sheathing is to be contained by posts).

Framing lumber and sheathing side posts

 Advantages:

- Low cost, widely available materials.
- Variations in size and dimension possible with same basic materials.
- Insulated portion prevents thermal bridge of solid material base plate.
- Structural capacity can be achieved with appropriate design.

 Disadvantages:

- Multiple components to be cut, sized and fastened results in more labor time.
- Mix of materials requires proper attention to sealing, fastening and connecting to prevent air leakage.
- Sheet materials may contain toxic adhesives and may off gas.

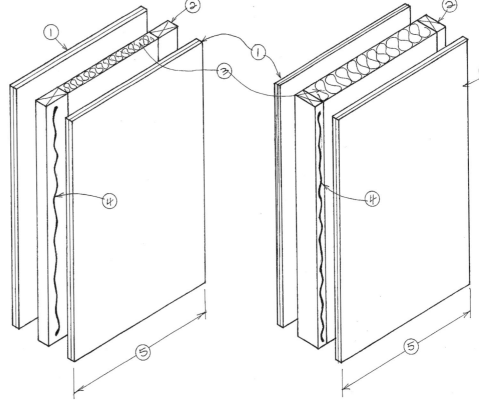

1. • Sheet materials typically ½" (12.5 mm); can be thicker if required for structural purposes or thinner to save on materials.
   • Inner sheet may not be required; bales can be installed directly against insulation and framing lumber.

2. • Framing lumber is typically 2×4 (38×89 mm).
   • Framing lumber laid flat is typical to create narrow side/post.
   • Framing lumber on edge can provide additional structural capacity and more depth of insulation.

3. • Insulation can include cellulose, varieties of batt insulation (including cotton, hemp and wool), straw, mineral wool, expanded clay or glass balls or perlite.
   • Cavity should be filled completely.

4. • All joints between materials to be caulked or glued to prevent air and moisture leakage.
   • Fastener size and frequency to be determined by structural requirements.

5. • Width of side/post to match bale width if plaster or sheathing is intended to cover the frame edge.
   • Width of side/post to match full wall width if plaster or sheathing is to be flush with frame.

I-joist side post

 Advantages:

- Low cost, widely available materials.
- Single component reduces labor input.
- Standard overall widths can match well with bale dimensions.
- Can be ordered to custom lengths, eliminating waste.
- Shape of joist can fit the ends of bales laid on edge, eliminating the need for additional insulation (see illustration).
- Insulation blocks can be inserted to provide positive locking during installation.
- Structural capacity can be achieved with appropriate design.
- Can be used with any type of base plate and top plate.

 Disadvantages:

- OSB web creates thermal bridge.
- OSB web may contain toxic adhesives and may off gas.
- Off-list use of product; structural analysis required to establish load-bearing capacity.

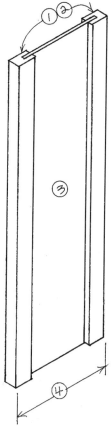

1. Flanges come in a range of widths from 1⅜" (35 mm) to 2½" (63.5 mm) and depths of 1¾" (44.5 mm) to 3½" (89 mm).

2. Flanges can be framing lumber or LVL.

3. Web made of OSB (oriented strand board) in thickness of ⅜" (9.5 mm), ⁷⁄₁₆" (11 mm) or ½" (12.5 mm).

4. Standard overall widths include 14" (355 mm) and 16" (406 mm).

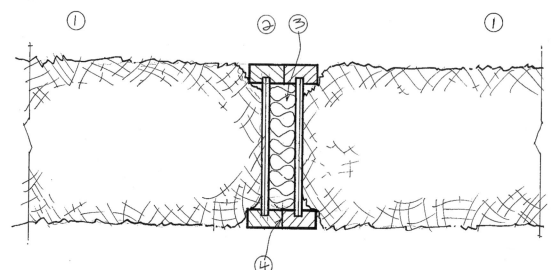

1. Straw bale.

2. I-joist.

3. Solid insulation (wood fiber, rigid mineral wool, or other).

4. Caulking or expanding tape to create seal.

Solid material side post

 Advantages:

- Single material minimizes labor input.
- Wood-wool or wood-fiber versions offer good thermal performance with no thermal bridges.
- Timber versions provide high structural capacity.

 Disadvantages:

- Materials must be special ordered and/or custom formed.
- Materials cost more than framing lumber options.
- Timber versions create a thermal bridge through the wall.
- Adhesives can contain toxic chemicals and may off gas.

1. • Glulam base plate made from solid framing lumber attached with adhesive.
   • Dimension of lumber dictates sizing.

2. • Laminated veneer lumber (LVL), parallel strand lumber (PSL), laminated strand lumber (LSL) or oriented strand lumber (OSL) made from wood plies or wood strands bonded with adhesive.
   • Come in a variety of sizes and can be custom ordered to exact dimensions.
   • Appropriate moisture protection must be used between glulam plate and foundation.

3. • Cement-bonded wood wool or cement-bonded wood fiber products are available in sheet form and can be cut to size for base plates. These are the only insulated option for solid material frames.

4. • Width of base plate to match bale width if plaster or sheathing is intended to cover the frame edge.
   • Width of base plate to match full wall width if plaster or sheathing is to be flush with frame.

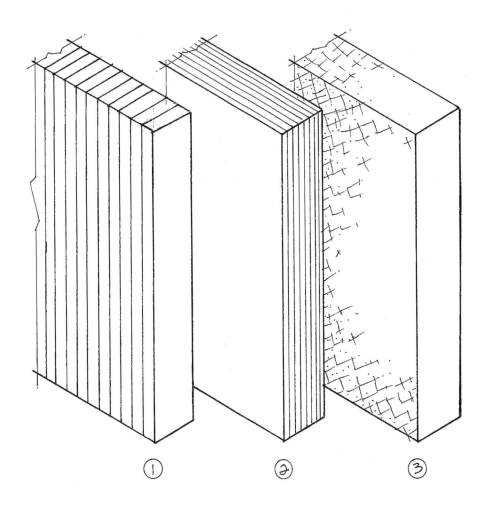

Top plates

Structural requirements

- Transferring loads to the walls and/or posts.
- Providing structural span over wall openings, as required.
- Connection points to the posts to resist all uplift forces.
- Connection points between panels, if necessary.
- Attachment points for structural sheathing, if used.
- Handling uplift forces during lifting and placement if the lifting force is applied at the bottom of the wall.

Building science requirements

- Thermal break from interior to exterior. If required to meet performance targets, the side plate/post may need to incorporate an insulation material or be made from a material with the required thermal resistance.

Functional requirements

- Providing attachment points and/or bearing surface for floor and roof framing, as required.
- Possible wiring chase. This can be built into the panel or external wiring chase baseboards may be secured to the base plate.
- Attachment points for decorative finishes on sheathing.
- Attachment point for trim and soffit, as required.

Framing lumber top plate

 Advantages:

- Low cost, widely available materials.
- Variations in size and dimension possible with same basic materials.
- Insulated portion prevents thermal bridge of solid material base plate.
- Structural capacity can be achieved with appropriate design.

 Disadvantages:

- Multiple components to be cut, sized and fastened results in more labor time.
- Mix of materials requires proper attention to sealing, fastening and connecting to prevent air leakage.
- Sheet materials may contain toxic adhesives and may off gas.

1. • Sheet materials typically ½″ (12.5 mm); can be thicker if required for structural purposes or thinner to save on materials.
   • Top or bottom sheet may not be required; bales can be installed directly against insulation and framing lumber, or top can be left open if insulation can be exposed to weather.
2. • Framing lumber is typically 2x4 (38x89 mm).
   • Framing lumber is positioned and sized according to required structural capacity and/or fastening of floor or ceiling joists.
   • I-joist can be substituted for framing lumber if additional structural capacity required.
3. • Insulation can include cellulose, varieties of batt insulation (including cotton, hemp and wool), straw, mineral wool, expanded clay or glass balls or perlite.
   • Cavity should be filled completely.
4. • All joints between materials to be caulked or glued to prevent air and moisture leakage.

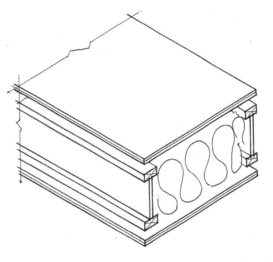

I-joist substituted for framing lumber.

- Fastener size and frequency to be determined by structural requirements.
5. • Width of top plate to match bale width if plaster or sheathing is intended to cover the frame edge.
   • Width of top plate to match full wall width if plaster or sheathing is to be flush with frame.

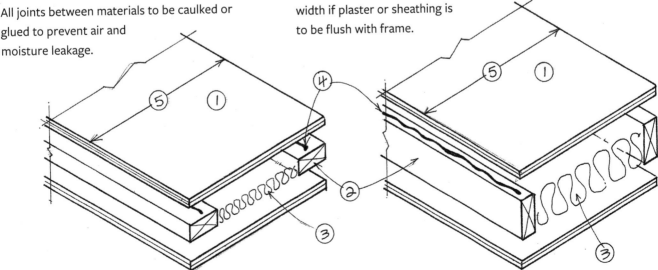

Solid material top plate

 Advantages:

- Single material minimizes labor input.
- Wood-wool or wood-fiber versions offer good thermal performance with no thermal bridges.
- High structural capacity from wood options.

 Disadvantages:

- Materials must be special ordered and/or custom formed.
- Materials cost more than framing lumber options.
- Solid wood materials create a thermal bridge through the wall.
- Adhesives can contain toxic chemicals and may off gas.

1. • Glulam top plate made from solid framing lumber attached with adhesive.
   • Dimension of lumber dictates sizing.

2. • Cement-bonded wood wool or cement-bonded wood fiber products are available in sheet form and can be cut to size for top plates. These are the only insulated option for solid material frames.

3. • Laminated veneer lumber (LVL), parallel strand lumber (PSL), laminated strand lumber (LSL) or oriented strand lumber (OSL) made from wood plies or wood strands bonded with adhesive.
   • Come in a variety of sizes and can be custom ordered to exact dimensions.

4. • Width of top plate to match bale width if plaster or sheathing is intended to cover the frame edge.
   • Width of top plate to match full wall width if plaster or sheathing is to be flush with frame.

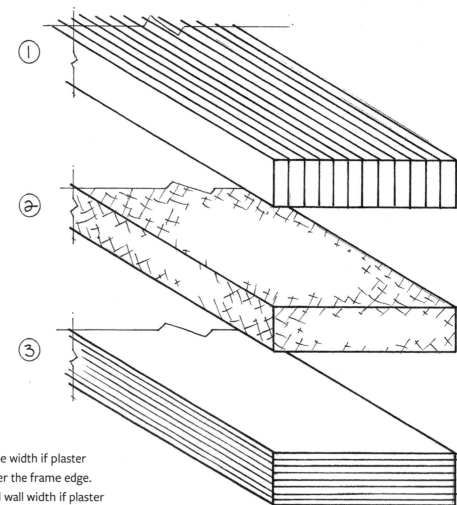

Creating hybrid frames

Most builders of prefabricated straw bale wall panels build frames with base, sides and top plates built from the same materials. However, different base, side and top plates can be combined to meet specific goals. As long as connection and sealing details are appropriate between unlike materials, none of the options outlined above are incompatible with one another.

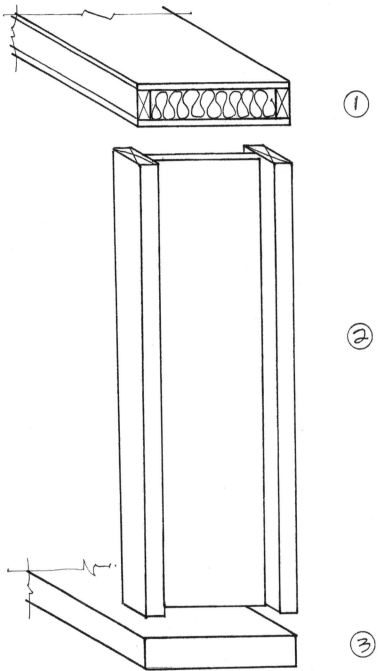

This is one option for creating a hybrid frame system:

1. Framing lumber top plate. Offers low cost, widely available materials, good thermal properties, good structural properties.

2. I-joist side post. Offers low cost, widely available materials, one-piece construction.

3. Cement-bonded wood wool base plate. Offers good thermal properties and high degree of moisture resistance.

Frame dimensions

Frame dimensions are typically chosen to match either the size of the straw bale (allowing the plaster or sheathing to cover the frame) or the size of the finished wall (allowing the frame to be exposed).

Frame covered

 Advantages:

- Allows for seams between "like" materials; easier to blend seams after installation.

- Simpler air sealing details. Plaster or sheathing can be sealed to frame, and single joint at seams easier to get tight.

- Less potential for water infiltration at bottom edge of frame.

- Standard size panels allow for use of sheathing material with low labor and no waste.

 Disadvantages:

- Plaster skin will need reinforcing over frame, may be prone to chipping or breaking.

- Attachment of panels during installation limited, as frame sides are not accessible for fasteners.

- Frames can only transfer structural loads to plaster or sheathing through fasteners.

1. Interior plaster skin or permeable sheathing covers frame.

2. Frame (of any material) same width as bales.

3. Exterior plaster skin or permeable sheathing covers frame.

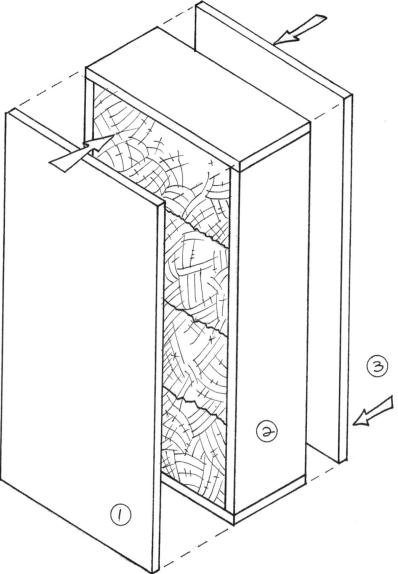

Frame exposed

 Advantages:

- Edge of frame can be used to screed plaster flat and smooth.
- Exposed frame sides easy to attach together.
- Frames can transfer structural loads to plaster or sheathing, especially shear loads.
- Plaster skins less prone to damage around edges.

 Disadvantages:

- More work and material required to cover frames after installation.
- Plaster joints over frame more prone to cracking.
- Fitting of sheathing material inside frame can be more labor intensive/wasteful unless sides match sheathing.

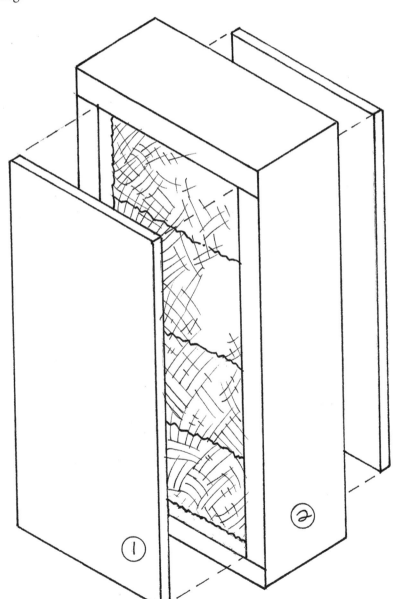

1. Interior plaster skin or permeable sheathing fits inside frame.

2. Frame (of any material) wider than bales to align with sheathing.

3. Exterior plaster skin or permeable sheathing fits inside frame.

Hybrid frame widths

Frames can be built with top, sides and bottom of different widths to make best use of the inherent advantages to suit the goals of the project.

 Advantages:

- Sheathing materials form joint between panels, no need to bridge over framing.
- Structural loads from top plate can be transferred into sheathing.
- Top and bottom plates exposed for fastening trim.

 Disadvantages:

- Additional variables for frame builder.
- Can be more difficult to provide air sealing.

1. Interior plaster skin or permeable sheathing covers frame sides and aligns with top and bottom framing.

2. Frame (of any material) with sides the same width as bales and top and bottom wider.

3. Exterior plaster skin or permeable sheathing covers frame sides and aligns with top and bottom framing.

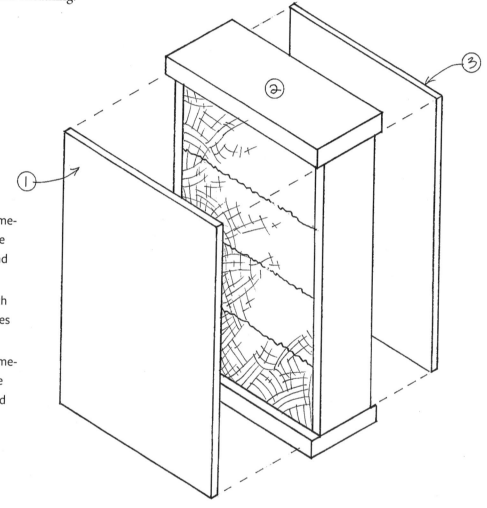

Bracing for Frames

Depending on the type of plaster or sheathing being used and the structural loads imposed on the wall panel, the panels may require some internal bracing. Bracing may necessitate channeling the straw bales to allow the braces to sit flush with the surface of the bale, and this can add significant labor time. Be sure to analyze the structural needs before choosing to use bracing.

Horizontal or vertical bracing

Horizontal bracing can help keep the long sides of the frame tied together and prevent distortion from the force of the bales pressing against the frame. Sheathing can be fastened at the midpoint in the wall.

Vertical bracing offers a fastening surface to sheathing similar to a stud frame wall, allowing conventional nailing/screwing pattern and load path analysis.

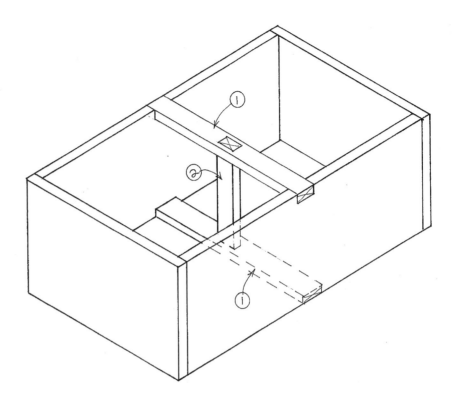

1. 1x4 or similar let into sides or top and bottom of frame.

2. Optional brace to join both horizontal braces across the wall.

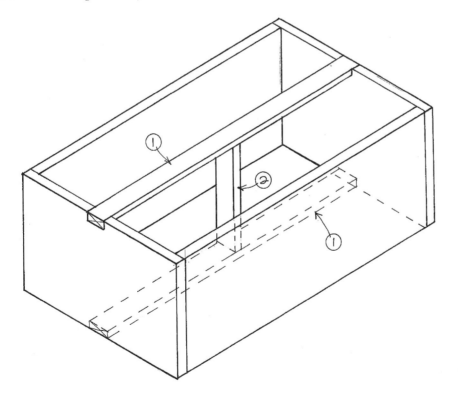

Diagonal bracing

Diagonal bracing can provide significant racking or shear strength to the frame. This type of bracing can require a lot of labor to cut and fit. Though it may offer fastening points for sheathing, it may be difficult to fit the bracing accurately.

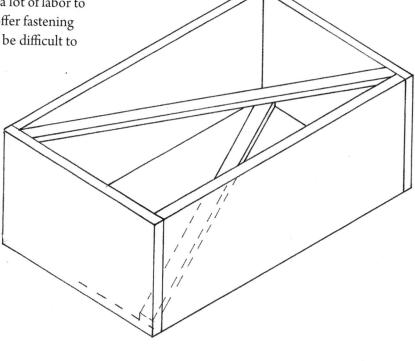

Buried horizontal brace

This bracing is installed between courses of bales to prevent the sides of the frame from bowing under pressure from tightly installed bales.

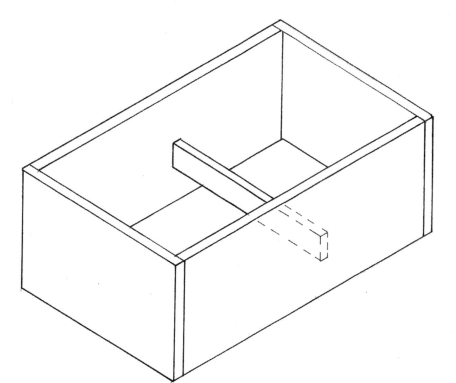

Frame Connections

Once the designer/builder has chosen the frame components, the connections between each component will need to be decided. The variety of frame components creates an array of potential connection types and fasteners.

Connections and fasteners must be selected to handle structural loads and meet production goals. Connections between framing components fall into three basic categories: *butt joints, tongue joints* and *lap joints.*

Butt joints

This type of connection is the most straightforward to accomplish. Any one of the fastening options shown may be used, or they may be used together.

The use of adequately sized structural screws is the fastest and least expensive means of connection, but is not as structurally robust as metal straps or L-brackets. L-brackets or straps and fasteners used on the sides of a panel may interfere with panel-to-panel connections. Those used on the face may interfere with sheathing connections.

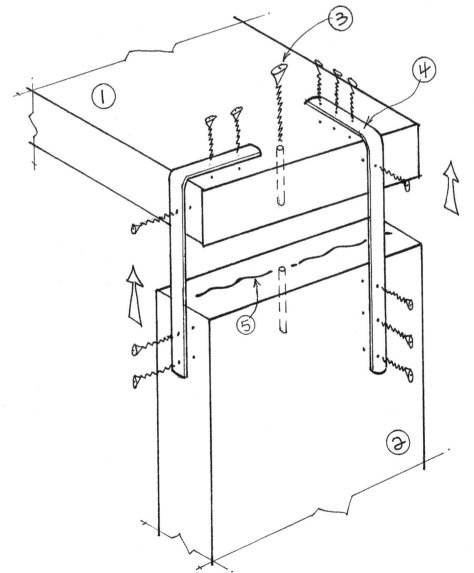

1. Top plate (or base plate).
2. Side post.
3. Structural screw (quantity and dimensions determined by structural analysis).
4. Metal strap or L-bracket (quantity and dimensions determined by structural analysis), can be on face and/or sides of frame.
5. Adhesive or caulking, as required.

Tongue joints

Tongue joint connections require more planning and fabrication time, but are inherently stronger and/or provide for better fastening.

Any one of the connection options shown may be used, or they may be used in conjunction.

The tongue joint is most applicable for solid-material frame components, although framing lumber elements can also use this joint.

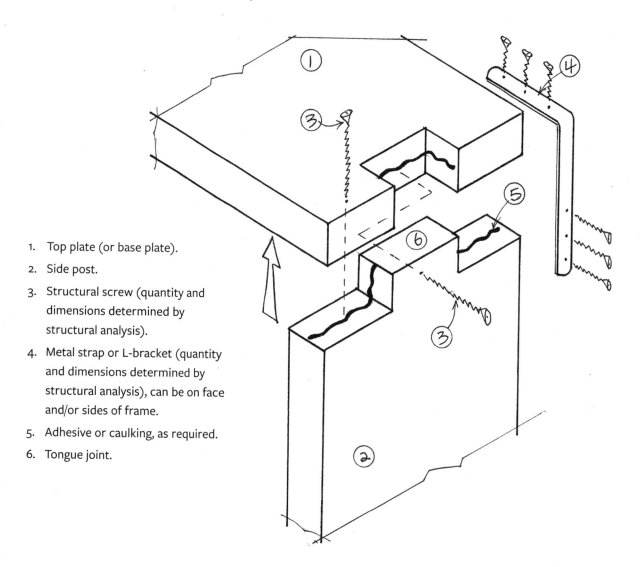

1. Top plate (or base plate).

2. Side post.

3. Structural screw (quantity and dimensions determined by structural analysis).

4. Metal strap or L-bracket (quantity and dimensions determined by structural analysis), can be on face and/or sides of frame.

5. Adhesive or caulking, as required.

6. Tongue joint.

Lap joints

The lap joint is most applicable for framing lumber and sheathing elements, where one piece of sheathing can be longer than the framing and create a positive lap with the adjoining element.

Example A relies on the horizontal structural screw to handle live and dead loads imposed at the joint, as the sheathing overlap will not likely provide adequate structural support. However, this example provides weather protection for the post element below, especially if the post incorporates insulation materials that may be damaged if wet.

Example B places structural loads directly on the post, and the vertical structural screw provides the main connection. This joint may not provide complete weather protection for the post, if required to keep insulation materials dry.

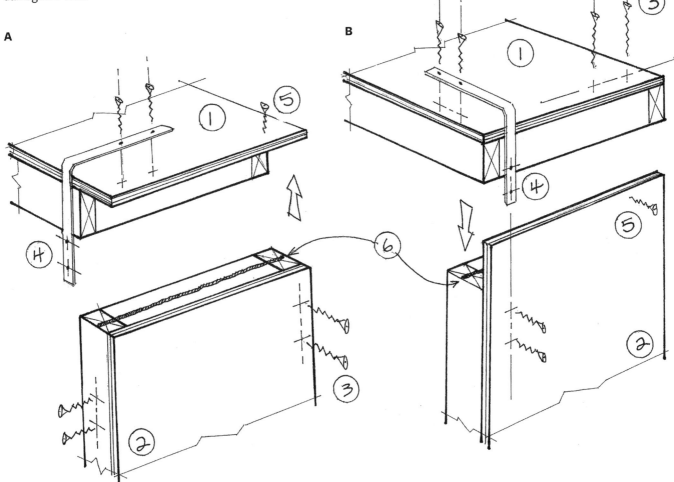

1. Top plate (or base plate)

2. Side post.

3. Structural screw (quantity and dimensions determined by structural analysis). Structural screw should connect through greatest thickness of plates.

4. Metal strap or L-bracket (quantity and dimensions determined by structural analysis), can be on face and/or sides of frame.

5. Screw or nail to secure sheathing only, not a structural connection.

6. Adhesive or caulking as required.

Sheathing Connections
Wet process, plaster covers frame

Plaster skins can cover the framing, providing plaster-to-plaster joints between panels.

Advantages:

- Thinner joints between panels can be easier to fill and cover with finish plaster.
- Simpler and better air sealing details between panels.

Disadvantages:

- Additional step of creating and installing temporary formwork around frame.
- Additional materials and steps for preparing and adhering mesh.
- Frame does not pass structural loads directly into plaster skins, relies on mesh connections.
- Plaster edges more prone to chipping and breakage during handling.
- Frames cannot be connected with fasteners along the side frames without penetrating plaster.
- Plaster requires curing time before panel can be moved/installed.

1. Frame component (any style).

2. Straw bales.

3. Temporary plaster form around edge of panel. This is set to the desired thickness of plaster, typically 1 inch (25 mm) and detached from the frame after plaster is cured.

4. Mesh covering framing and extending onto the surface of the straw bales by 2–6 inches (50–152.5 mm). Mesh can be metal, plastic or fiberglass.

5. U-pins to fasten mesh into straw bales.

6. Fasteners to attach mesh to frame. Can be screws with washers, staples or other fastener to meet structural requirements.

Wet process, plaster flush with frame

Plaster skins can be flush with the frames, resulting in wood-to-wood connections between panels.

 Advantages:

- Frame can be used as a screed to level the plaster.
- Exposed frames allow for easy panel-to-panel fastening along the posts.
- Structural loads can be transferred from the frame to the plaster skins, potentially reducing the loads on the frame and therefore the size of the framing members.
- Frames can be left exposed or trimmed as part of wall system aesthetic.

 Disadvantages:

- Additional materials and steps for preparing and adhering mesh.
- Air sealing can be difficult around the plaster-frame perimeter.
- Finish coat plaster (if being used) must bridge all framing, requiring extensive application of mesh.
- Plaster requires curing time before panel can be moved/installed.

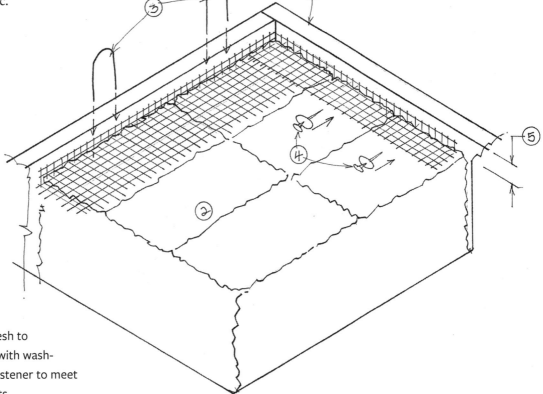

1. Frame component.
2. Straw bales.
3. U-pins to fasten mesh into straw bales.
4. Fasteners to attach mesh to frame. Can be screws with washers, staples or other fastener to meet structural requirements.
5. Frame extends beyond face of bales by desired thickness of plaster, typically 1 inch (25 mm).

Dry process, sheathing covers frame

Sheathing can be attached directly to the face of the frame.

 Advantages:

- Panel frames can be constructed to match conventional sheathing sizes, resulting in efficient use of materials.
- Sheathing can be used to keep frame square.
- Air sealing is easy to achieve by use of adhesive or caulking between frame and sheathing.

- Joints between panels simple to seal.
- With adequately sized and placed fasteners, sheathing can take some degree of structural loads.

 Disadvantages:

- Sheathing hides frames, making post-to-post connections more difficult.
- Structural loads not transmitted directly to sheathing, reliance on fasteners.

1. Frame component.

2. Straw bales flush with frame.

3. Permeable sheathing.

4. Fasteners, as required. If sheathing is providing structural role, fasteners must be sized and positioned accordingly.

5. Adhesive or caulking as required for structural and air-sealing purposes.

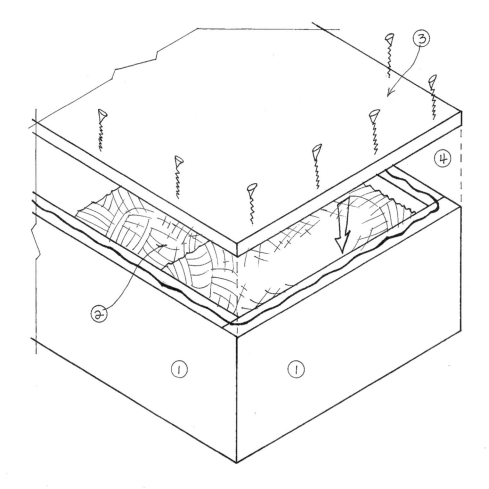

Dry process, sheathing inside frame

Sheathing can be fit inside the frame to finish flush with the face of the frame.

 Advantages:

- Panel frames can be constructed to match conventional sheathing sizes, resulting in efficient use of materials.
- Exposed frames allow for easy panel-to-panel fastening along the posts.
- Structural loads can be transferred from the frame to the plaster skins, potentially reducing the loads on the frame and therefore the size of the framing members.
- Frames can be left exposed or trimmed as part of wall system aesthetic.

 Disadvantages:

- Not all sheathing types can accept fasteners along their narrow edge.
- Can be more difficult to air seal along joints between sheathing and frame. Custom-sized panels require very accurate cuts along sheathing to maintain tight fit with frame.

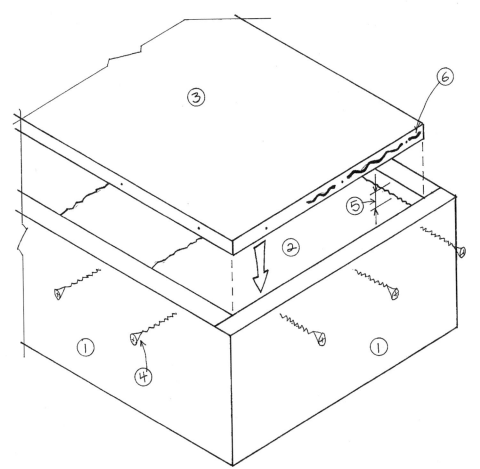

1. Frame components.
2. Straw bales.
3. Permeable sheathing.
4. Fasteners as required. If sheathing is providing structural role, fasteners must be sized and positioned accordingly.
5. Frame extends beyond face of bales by thickness of sheathing, typically from ¾–2 inch (19–51 mm).
6. Adhesive or caulking as required.

Hybrid process, sheathing covers frame

The hybrid process combines a plaster coating on the straw bales with a permeable sheathing material.

Advantages:

- Thinner and less expensive sheathing and plaster can be used.
- Interior joints between panels can be mudded and taped, rather than requiring skim coat of plaster.

- Combined structural system of both plaster and sheathing.

Disadvantages:

- Double labor input for both plastering and sheathing processes.
- Sheathing hides frames, making post-to-post connections more difficult.
- Plaster requires curing time before panel can be moved/installed.

1. Frame component.

2. Straw bales.

3. Plaster applied to straw bale, screeded level with face of frame.

4. Permeable sheathing, typically ½ inch (12.5 mm).

5. Frame extends beyond face of bales by thickness of plaster, typically ½ inch (12.5 mm).

6. Adhesive or caulking as required.

7. Fasteners as required. If sheathing is providing structural role, fasteners must be sized and positioned accordingly.

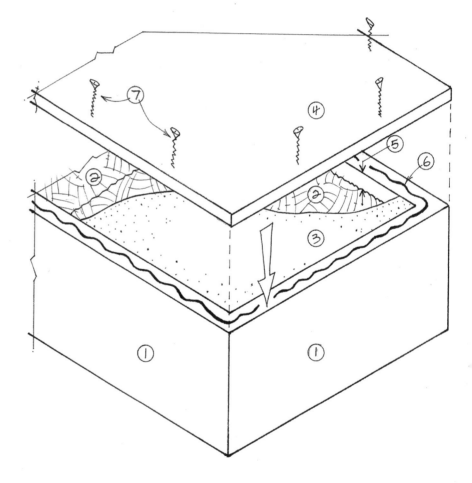

Panel-to-panel Joints: Internal Connection Elements

To maximize the thermal performance of the walls, the seams between panels must be designed and built to minimize thermal bridges and air leakage. The design of the joints must ensure that the installer is able to accurately complete the work required on site in an environment that considers site conditions such as weather and the pressures of time during installation.

There are a few variations on panel-to-panel joints that may be employed. All of these involve the use of some form of caulking, adhesive or expanding tape to create a positive seal between panels. The selection and application of these products will have an important impact on the performance of the walls so care should be taken when making a decision.

Butted seam

The square ends of the panel are caulked or taped and then pressed together. This can work with most frame styles and is the simplest joint to manage during installation. It relies fully on the adhesive material to seal the joint, unless further layers of plaster, sheathing or house wrap air barrier will also be used to add a degree of air tightness.

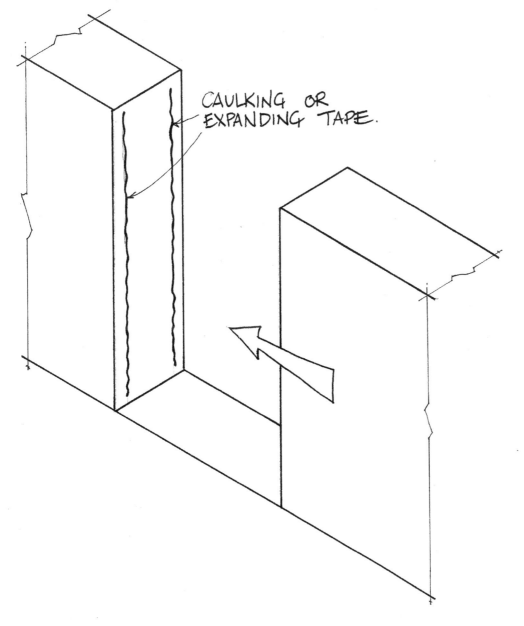

CAULKING OR EXPANDING TAPE.

Keyway seam

The keyway seam on the right requires a channel to be designed into the side panel, and will work best with solid-material side plates. The seam below is achieved by using the natural shape of the TJI side plate or by leaving the sheathing material off of the exposed side of a framing lumber side plate.

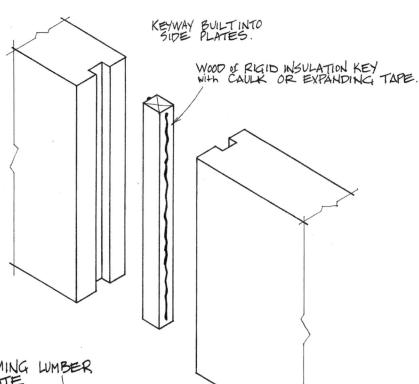

KEYWAY BUILT INTO SIDE PLATES.

WOOD or RIGID INSULATION KEY with CAULK OR EXPANDING TAPE.

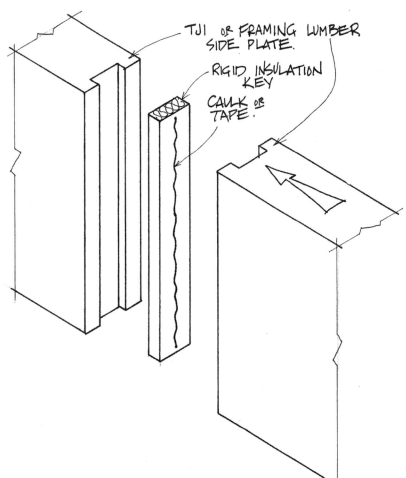

TJI or FRAMING LUMBER SIDE PLATE.

RIGID INSULATION KEY

CAULK or TAPE.

A key made from framing lumber or rigid insulation can be inserted into the keyway on the fastened panel and will slot into the keyway on the next panel installed. The panel being installed will need to move sideways to fit around the key, or the key may be able to be installed once both panels are in place if sideways movement cannot be achieved. Caulking or expanding tape on the key will provide the final air seal. This system can be more labor intensive to build and install, but provides positive connection from panel to panel.

Lapped seam

The lapped seam can be created in the manufacture of the side plate, or framing lumber or rigid insulation can be affixed during installation. If the lapped seam is created when the panel is built, care will need to be taken to ensure proper lapping order when the panels are installed.

Caulking or expanding tape can be applied to the matching faces of the lap joint and/or the butted ends of the panels. Lap joints will need to be matched to door and window openings to ensure that framing is exposed where necessary to connect with openings.

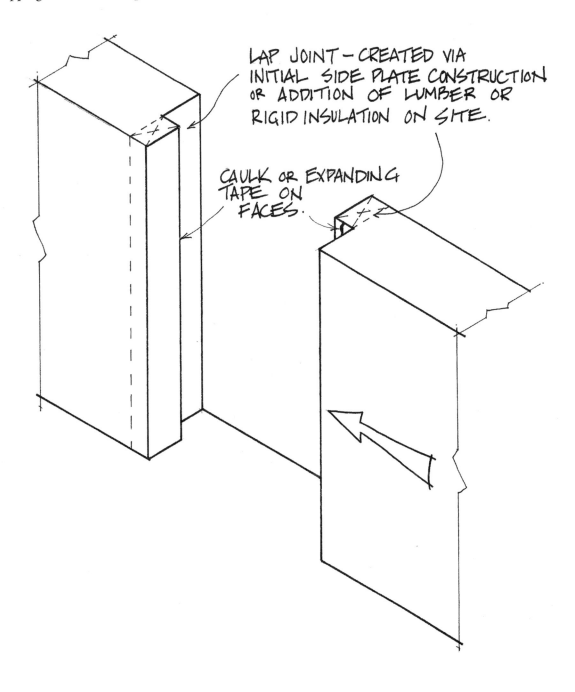

LAP JOINT — CREATED VIA INITIAL SIDE PLATE CONSTRUCTION OR ADDITION OF LUMBER OR RIGID INSULATION ON SITE.

CAULK OR EXPANDING TAPE ON FACES.

Panel-to-panel Joints: Exterior Connection Elements

The visible joints between panels can be handled in many ways, depending on the type of panel, the framing style and the chosen finishes for the exterior and interior of the wall. These types of joints can be mixed on the same panels, using one joint style on the interior and a different joint for the exterior.

Trim joint

When the plaster or sheathing on the panels is intended to be the finished wall surface on the interior and/or exterior, the joints can be finished with a face-mounted trim. The trim would be part of the overall trim plan aesthetic for the building. Trim can be wood, metal or plaster.

👍 *Advantages:*

- Speed and simplicity of installation.
- Wood or metal trims cover the seams with a material that will not telegraph any cracks or slight movement between the panels.

👎 *Disadvantages:*

- Panels must be shipped with a paint-ready surface and need to be adapted to the aesthetics of the building to suit the exposed trim.

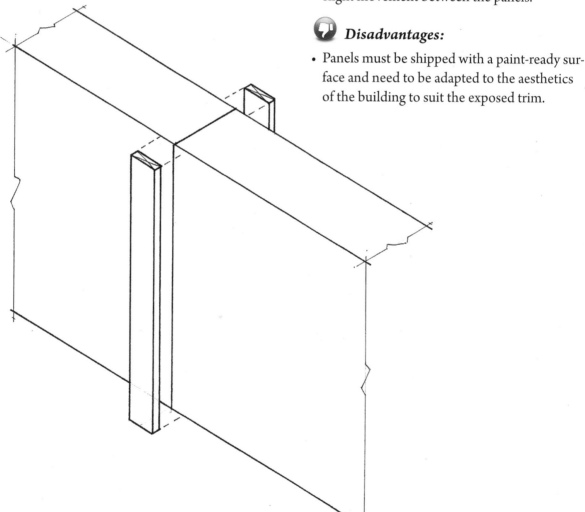

Sheathed joint

The sheathed joint is covered by a final layer of material. This could be a sheet material (including drywall, MgO board, or other permeable finished sheathing) or a final coat of plaster for wet-process panels.

 Advantages:

- Ease of installation for conventional sheet materials.
- Additional air control layer and final "seamless" finish.

 Disadvantages:

- Additional cost for materials and installation.
- For plastered finishes, the seams must be mudded and taped to help prevent cracks from telegraphing at joints.

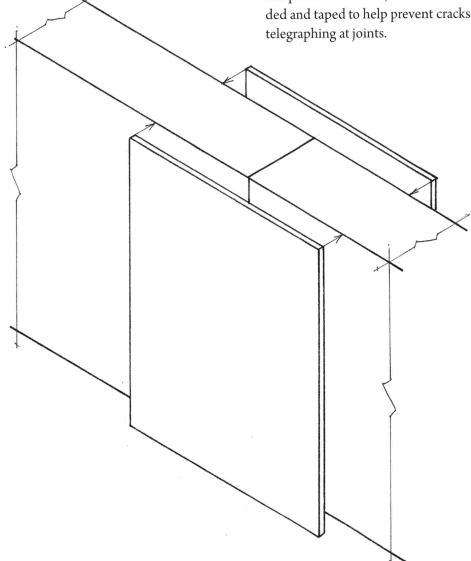

Furring and cladding joint

The furred and cladded joint employs strapping to create a fastening surface for a wide array of sheathing and cladding materials. The panel seams can be caulked or taped to provide air sealing.

 Advantages:

- A ventilated rain screen on the exterior gives additional rain protection and a raceway for electrical and/or plumbing runs on the interior.

- The strapping provides convenient attachment points for trim and interior cabinetry. Panels do not need to be built with a final finish surface.

 Disadvantages:

- Additional material and labor costs and an increased wall width.

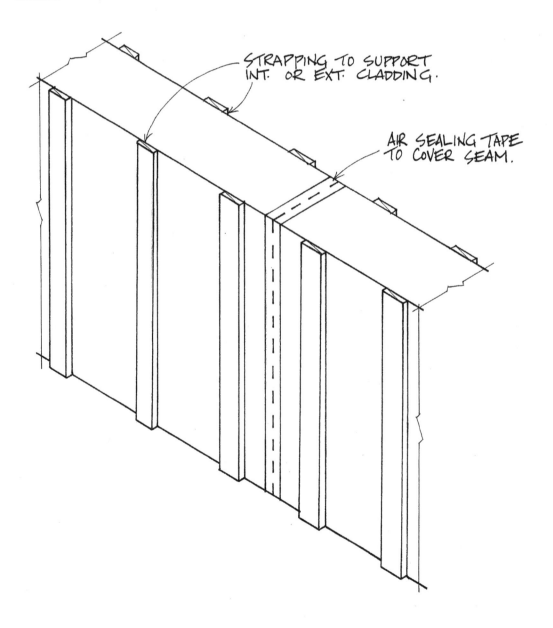

STRAPPING TO SUPPORT INT. OR EXT. CLADDING.

AIR SEALING TAPE TO COVER SEAM.

Door and Window Openings

There are two distinct ways of approaching door and window openings in S-SIPs. These approaches can be mixed within a single project. In particular, door openings are best handled with the first two options, as there is no need for an integrated sill.

Panel with site framed opening

Solid panels are built to the edges of each door and window opening, and a custom panel is built to accommodate the door or window framing. The panel with the opening integrates all the required framing for the opening and is sheathed in the same manner as the solid panels.

The furred and cladded joint employs strapping to create a fastening surface for a wide array of sheathing and cladding materials. The panel seams can be caulked or taped to provide air sealing.

S-SIPs are used to create all the solid wall sections of this building, and the window and door openings have been framed-in on site.

 Advantages:

- Fast installation times.
- No custom work or component assembly required on site.

 Disadvantages:

- Panels with openings are more fragile to handle and ship.
- Little or no leeway for errors in foundation size.

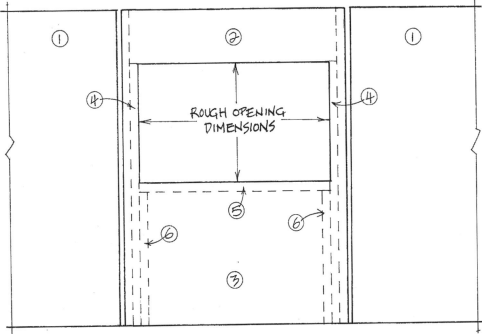

1. Solid panels.
2. Structural header (to meet code req's) integral to panel.
3. S-SIP section or framed and insulated *in situ.*
4. Framing to bear header loads integral to panel.
5. Sill framing.
6. Cripples to support sill.

Sill sections are created as S-SIPs, and the sides of the panels include bearing points for a structural header.

Component opening

Solid panels are built to the edges of each door and window opening, and custom components are assembled on site to create the opening section. The components can be prefabricated and delivered along with the walls, or created *in situ*. The framing used to support the header component is fastened directly to the adjacent solid panels, and the sill component is fitted as a distinct panel with the sill framing integrated.

 Advantages:

- Header and sill components less prone to shipping damage.
- *In situ* framing can be adjusted to compensate for foundation size issues.

 Disadvantages:

- More distinct components to build and ship.
- More labor input during installation.

1. Solid panel.
2. Structural header component.
3. Sill section component.
4. Framing to support header secured to solid panels.
5. Sill framing.
6. Sill support framing.

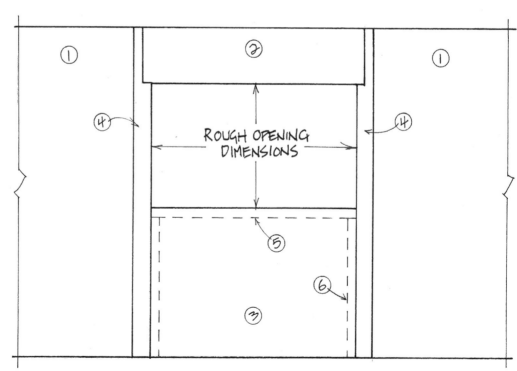

Openings built into panels

Window and door openings are integrated into the panel layout and built directly into the panel. In this approach, the panelization of the building is not dependent on the placement of openings; rather, the openings are placed into the panel where needed.

 Advantages:

- Faster installation times.

 Disadvantages:

- Custom placement and framing adds extra labor during production.
- Placement and dimensions must be accurate in production.
- Panels with openings are more fragile to ship and install.

Window openings can be incorporated into an S-SIP.
CREDIT:
BEN POLLEY

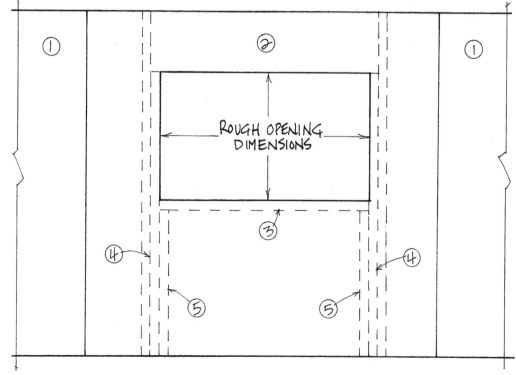

1. Solid panel.
2. Header integrated into panel.
3. Sill integrated into panel.
4. Header support.
5. Sill support.

Electrical Wiring

One of the main challenges with any type of prefabricated wall system is dealing with electrical wiring and other services that need to run inside or through the wall. As with any prefab wall, there are four basic ways that wiring can be run in prefab bale panels. A single system may be chosen for the entire project, or it may be beneficial to use a variety of approaches depending on the context.

Surface mounting of wiring and boxes

In its simplest and least expensive variation, this approach uses metal-shielded cable that is code-approved for surface mounting, or metal conduit to shield regular wires running to surface mounted electrical boxes. This approach is typical of many commercial/industrial applications, where wiring needs can change and where the aesthetics of surface-mounted wiring are not an issue.

Prefabricated surface-mounted wiring systems can be purchased from any electrical supply company, and include a wide range of styles, materials, and box/switch options. Whole-house systems will typically include a continuous baseboard channel (often with outlet receptacles built in) that replaces conventional trim boards, with surface-mounted branches that bring wires up to switches. Though there is a higher up-front cost for these systems, when compared against the cost of running conduit and/or wires through walls and making final connections later (and eliminating the cost and labor for baseboard trim), it may be minimal. Surface-mounted wiring has the added bonus of being very easy to change or modify in the future — great for building resilience.

Air tightness of the wall is very easy to achieve with this system, as the air barrier may not need to be punctured at all. If fasteners for the surface-mounted wiring system do penetrate the plaster or sheathing, the holes can be dabbed with caulking before screws are placed to minimize penetrations into the wall.

For the purposes of simplifying building prefabricated wall panels, surface-mounted wiring is ideal if the cost and aesthetic considerations make this approach feasible. It should be noted, however, that baseboard-mounted outlets may not meet accessibility standards in some jurisdictions.

Wiring at panel joints

Some builders of prefab bale panels have chosen to plan for wiring and electrical boxes to run at the seams between panels and/or between panels and door or window framing. This can be achieved by building raceways into the panel framing, or by leaving spaces between the panels and adjacent framing where wires and boxes can be installed.

This type of approach can work well with a base plate design that also accommodates wiring.

This option works best with framing lumber posts, where the framing lumber can be set back from the face of the wall while the sheathing material can be used as a plaster screed in wet-process panels or to fasten structural sheathing.

Wiring runs can enter the raceways from below the floor or from the ceiling with equal ease, and the wooden elements of the frame can be used to anchor wires. The wiring chases can be covered after wall and electrical installation on site, with sheathing or plaster backing. The final finish on the walls will cover these seams. This approach can limit the placement of switches and outlets in the building, and will require coordination of electrical needs at the planning stages of the building.

It is relatively easy to maintain an effective barrier plane using this strategy, as the raceway can be effectively sealed without relying on the cover or fill to play a role in keeping the wall airtight. However, the planning for this approach must take into account code requirements that may dictate the allowable spacing between outlets, and may necessitate panel joints where they would not otherwise be needed.

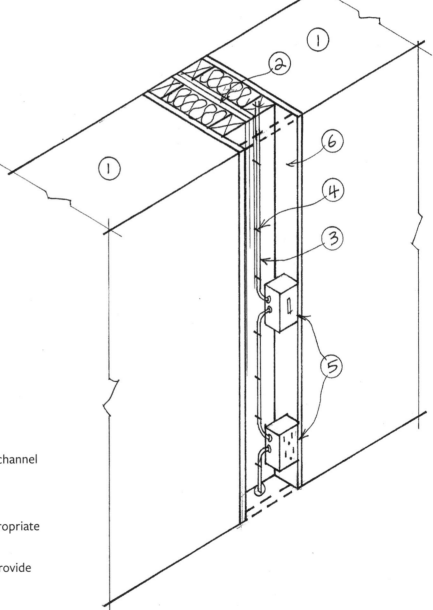

1. Panel
2. Side posts built to create wiring channel
3. Electrical wire
4. Wire staples attached to frame
5. Electrical boxes mounted to appropriate depth
6. Side post sheathing extends to provide fastening for sheathing

Routered wire pathways

This approach is common with many SIP and ICF systems, where the wall system is constructed and then the wiring layout is drawn onto the wall, and channels for wiring runs and boxes are cut or routered into the wall surface. Once all the wiring has been placed, the channels are filled and/or covered prior to the installation of the final wall finish.

For wet-process walls, cutting channels will be a dusty procedure, but plaster can be cut cleanly and accurately with a masonry-cutting wheel in a circular saw or hand-held grinder. Channels can be filled with plaster mix after the wiring is in place, keeping in mind that electrical codes may mandate that any plaster containing lime or cement cannot touch unshielded wiring.

Channels can be filled with clay or gypsum plaster over unshielded wiring.

In dry-process scenarios, the ease and effectiveness of creating and sealing channels will depend on the material choice and the thickness of the material. If a final finish is being applied over the structural sheathing, the channels may not need to be filled.

This approach can offer challenges where air tightness is concerned, as the channels are likely to be breaches in the main air barrier for the wall surface (plaster or structural sheathing). If the walls are covered with a finish plaster or a final layer of sheathing, it may be possible to plan on using this layer as the main air barrier.

1. Panel
2. Plaster or sheathing
3. Routered channel for wires
4. Routered opening for electrical boxes
5. Wire in channel
6. Air tight boxes mounted in plaster or sheathing

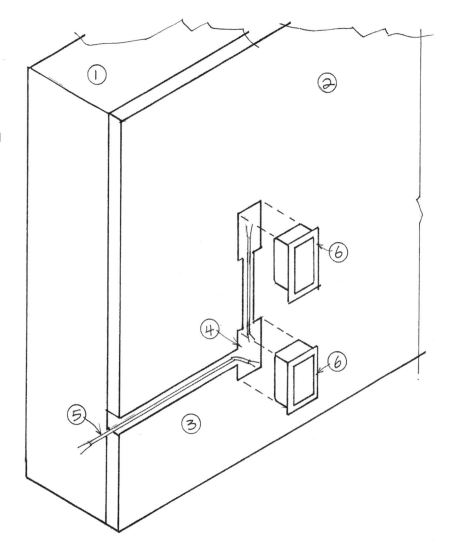

Embedded conduit

A suitable electrical conduit can be placed into the panels prior to sheathing while they are being assembled, providing a predetermined pathway for wiring that can be fished through the buried conduit after the walls are assembled on site.

Custom conduit layout

With this approach, conduit can be laid out to match a supplied wiring diagram for the building. If the conduit is custom built into the panels, it is possible to place all electrical boxes at the same time; this provides a system that is ready to wire on site.

The straw is channeled to accommodate conduit and boxes for a customized electrical installation.

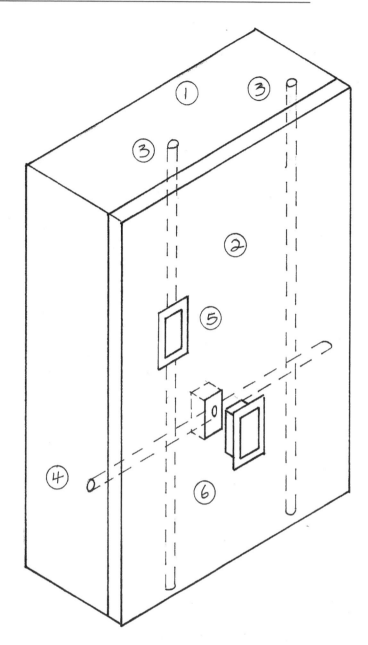

1. Panel
2. Plaster or sheathing
3. Vertical conduit, allows wire to be introduced from top or bottom of wall
4. Horizontal conduit, allows circuits to run continuously around room
5. Routered opening cut into conduit
6. Air tight electrical box

Regularly spaced conduit layout

Conduit can most easily be placed into the walls vertically, where access could be from either the top or bottom of the wall. Horizontal conduit runs are also possible, allowing for wiring to be run between panels. The placement of horizontal conduit will need to be very accurate to ensure it lines up between panels. Junction boxes can be cut into the plaster or sheathing directly over the conduit after the walls have been installed.

It should be noted that in many jurisdictions, this approach requires an electrical inspector to approve the conduit and boxes before they are covered with sheathing or plaster.

Various types of metal and/or plastic conduit may be appropriate; be sure to check local codes. For wet-process panels using a lime or lime-cement plaster, the conduit will need to be approved for contact with this material. The depth of the conduit behind the finished surface of the wall will dictate whether or not the conduit needs to provide puncture resistance.

It is best, if using this approach, to eliminate the need to carve channels into the straw bales for conduit, as this can add a significant amount of labor time to the process.

A marking system can be used on the face of the panels to indicate the locations of conduit runs.

An advantage of this approach is the flexibility for revising the wiring scheme and adding junction boxes in the future without needing to damage the final finish.

If the conduit is entirely covered by the plaster or structural sheathing, maintaining air tightness will be a matter of ensuring that junction box penetrations are appropriately sealed.

Conduit connects to a box embedded in the wall.

Foundation Connections

Structural requirements

- Securing panels to foundation to resist all lateral and uplift forces.
- Transfer all imposed loads into the foundation as required.

Building science requirements

- Provide a moisture break between foundation and wall.
- Provide an airtight seal between foundation and wall.
- Prevent thermal bridging across the wall at the foundation junction.

Functional requirements

- Attachment system must not interfere with panel installation process.
- Attachment points must align with structural elements in the panel frame.

S-SIPs can work with any type of foundation. The width of the wall invites creative strategies, such as (below) double-wythe stem walls and load sharing between foundation wall and floor framing (above right) system. This type of approach can provide adequate structural properties while using materials more efficiently and often creating better thermal performance.

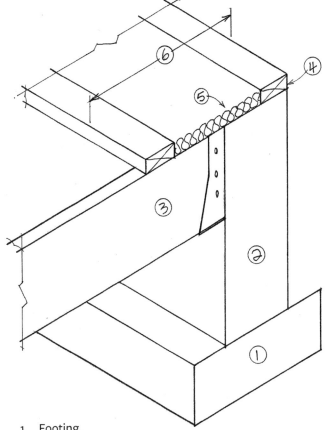

1. Footing.
2. Insulated structural wall (poured concrete, CMU, aerated concrete, compressed earth block or other suitable material).
3. Floor joist system hanging or ledgered to structural wall.
4. Sill plates fastened to foundation and to floor joists/floor sheathing.
5. Insulation between sill plates.
6. Width of between sill plates matches wall width.

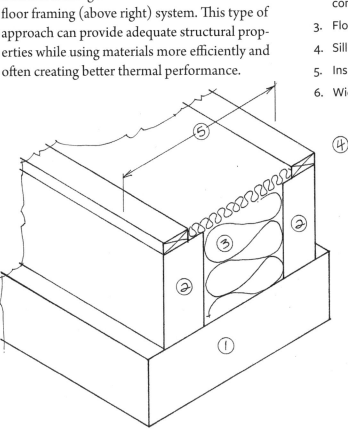

1. Footing.
2. Double-wythe structural wall (poured concrete, CMU, aerated concrete, compressed earth block other suitable material).
3. Insulated cavity.
4. Sill plates fastened on each wythe.
5. Width of foundation matches wall width.

Pre-installed sill plates to match wall width

Sill plates are pre-installed onto the foundation in line with the position of the walls. In this approach, a variety of different fastening options exist, depending on panel type and structural requirements.

1. Thickened edge slab or perimeter beam foundation, built to match width of wall.

2. S-SIP installed directly on top of pre-installed sill plates.

3. Sill plates fastened to foundation. Caulking or gaskets used to ensure airtight seal to foundation and to panel.

4. Anchor bolt or wedge anchor as required to fasten sill plate.

5. Structural toe-screw through base plate and into sill plate.

6. Metal plate or strap anchored to sill plate and/or foundation and panel base plate and/or side plate.

7. Metal L-bracket anchored to foundation and panel base plate and/or side plate.

8. Moisture-resistant insulation.

Pre-installed sill plates inside panel base plate framing

Sill plates are pre-installed onto the foundation to match the inside gap created by the base plate framing on the panel. Fasteners go through base plate framing and into sill plates.

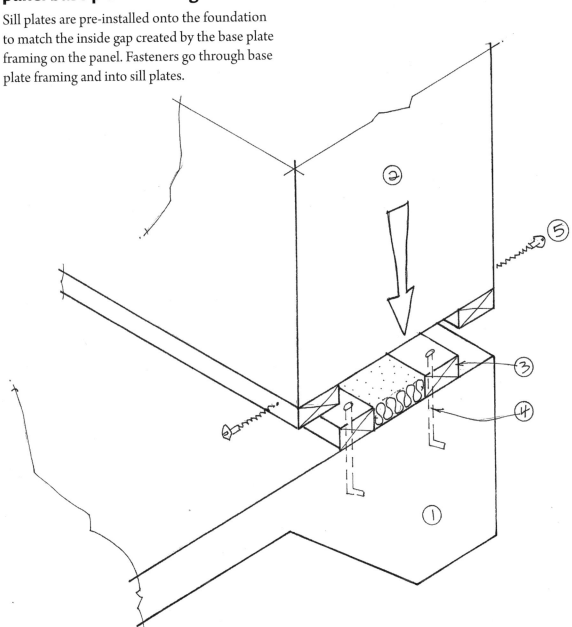

1. Thickened edge slab or perimeter beam foundation, built to match width of wall.

2. S-SIP installed so base plate framing straddles sill plates.

3. Sill plates fastened to foundation. Caulking or gaskets used to ensure airtight seal to foundation and to panel.

4. Anchor bolts or wedge anchors as required to fasten sill plates to foundation.

5. Structural screws connect panel base plate to sill plates as required.

Surface-mounted fasteners

No sill plates are pre-installed onto the foundation. The panel is placed directly on the foundation (using moisture resistant barriers where required, and providing caulking or other means of making an airtight seal) and a variety of different fasteners are used to attach the panel directly to the foundation.

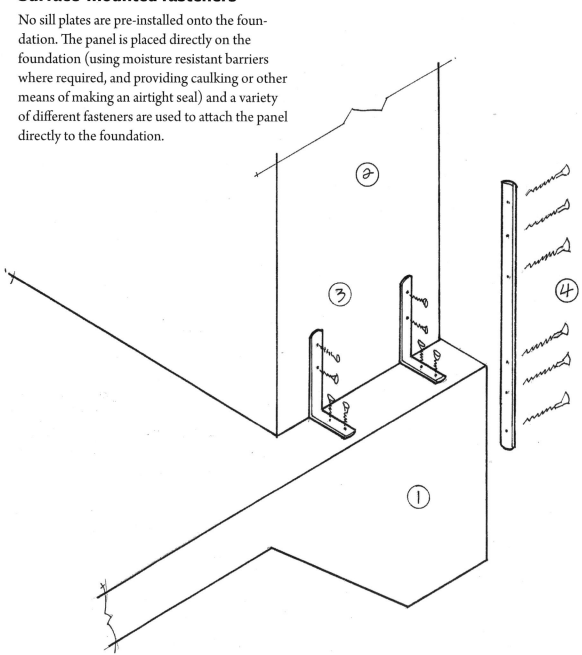

1. Thickened edge slab or perimeter beam foundation, built to match width of wall.

2. S-SIP placed directly on foundation over moisture resistant barrier, where required.

3. End-mounted fastener attached to foundation and base plate and/or side post.

4. Metal strap fastener attached to foundation and base plate and/or side post.

Floor and Roof Systems

Floor and roof systems require little or no adaptation in use with S-SIPs. Energy efficiency can dictate placement and attachment more than the wall configuration itself, and the extra width of the wall is a definite advantage when aiming for a high degree of energy efficiency.

Floor joists

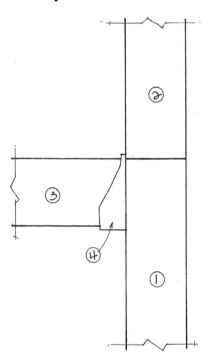

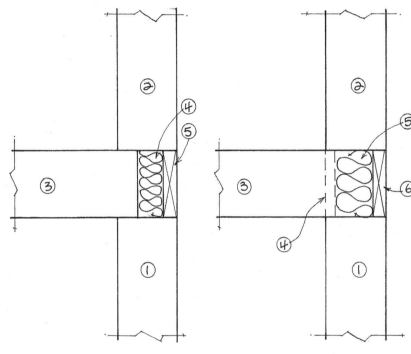

Joist hangs inside wall

1. Top of first story wall.
2. Bottom of second story wall.
3. Floor joist.
4. Top mounted floor joist hanger.

Connecting a floor system to a panel wall using hangers on the inside of the wall is the most energy efficient approach, maintaining a continuous layer of wall insulation and achieving ideal thermal properties and simple details for air tightness. However, the depth of the floor joist reduces the functional height of the wall, and the wall must be designed to account for this. Hanging joists on the interior simplifies the treatment of the exterior of the building by taking out the wide transition required between materials when joists rest on the wall.

The wall must be designed to be able to accept the loads from the floor joists on the inner edge.

Joist bears on top of wall, inner edge

1. Top of first story wall.
2. Bottom of second story wall.
3. Floor joist.
4. Insulation in cavity.
5. Rim joist.

This is a more conventional approach with the joist bearing directly on the wall, but it takes advantage of the width of the wall to provide a good deal of depth for insulation to minimize thermal bridging of the joists to the exterior.

The wall must be designed to accept the loads from the floor joists where they bear on the frame.

Joist bears on top of wall, outer edge

1. Top of first story wall.
2. Bottom of second story wall.
3. Floor joist.
4. Rim joist on inner edge of wall.
5. Insulation in cavity.
6. Rim joist on outer edge of wall.

This is the most conventional approach, and the least energy efficient. The voids created by the floor joists are difficult to insulate and air seal properly.

The loads from the floor joists are distributed evenly across the width of the wall.

Roof framing
Rafter on outside edge

Conventional roof rafters will be cut to bear on the outside edge of the panel. The panel must be designed to accept roof loads on the outside edge, or blocking must be provided to spread the load more evenly across the width of the wall.

This option bears the roof loads on the outside edge of the wall, and structural design must account for this eccentric loading. While this option allows for the most typical roof framing, it does not provide for adequate thermal protection at this junction. The addition of a small heel will help to balance loading and create space for additional insulation.

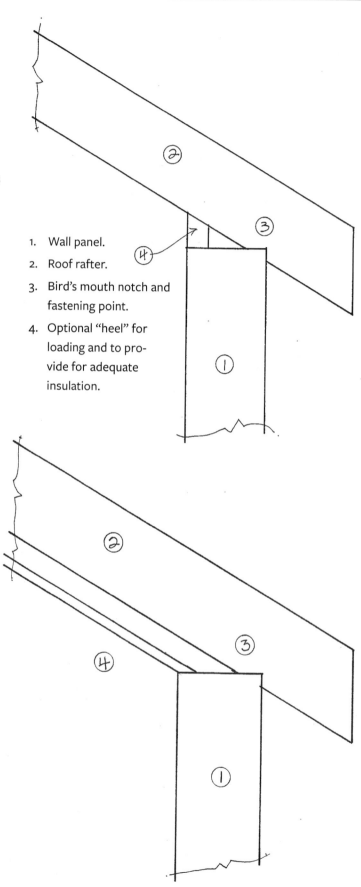

1. Wall panel.
2. Roof rafter.
3. Bird's mouth notch and fastening point.
4. Optional "heel" for loading and to provide for adequate insulation.

Rafter on outside edge, ceiling joist on inside edge

To provide extra depth for insulation, framing can be cut to bear on both edges of the wall panel.

1. Wall panel.
2. Roof rafter.
3. Bird's mouth notch and fastening point.
4. Ceiling joist on inside edge of wall.

Truss roof, typical

A truss roof can bear across the width of the wall. The designer can specify the loading point(s) according to the design of the walls.

1. Wall panel.
2. Truss ceiling member.
3. Truss rafter member.
4. Truss bearing point, to be determined by designer.
5. Truss fastening to wall to be determined by designer.

Truss roof, energy efficient

A truss roof can be designed with a "heel" that provides extra depth for insulation and allows the edges of the trusses to be thermally broken from the exterior.

1. Wall panel.
2. Truss ceiling member.
3. Truss rafter member.
4. Truss heel.
5. Rigid insulation.

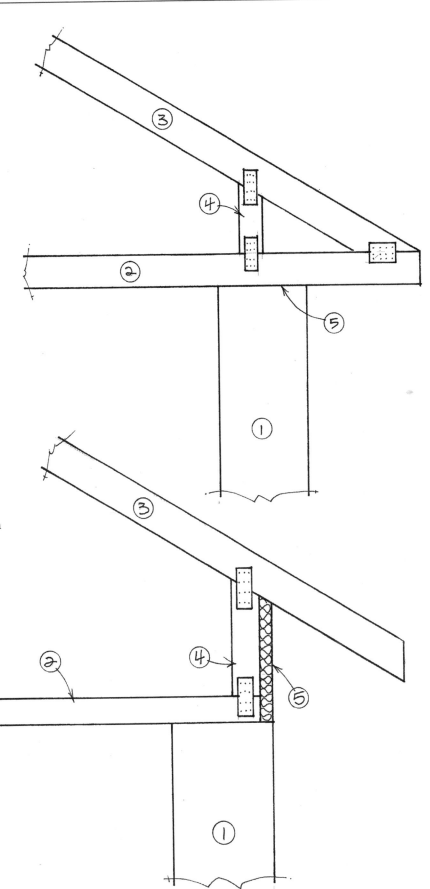

Chapter 7

Design Notes

A DESIGNER OF S-SIPs will need to choose from the wide range of options presented in the previous chapter in order to create a panel that works best for a particular project.

In addition, panelization requires a building designer to take many points into consideration at the design stage in order to realize the full extent of the labor and material savings offered by prefabrication. Understanding the standard material dimensions for the elements being used in the chosen S-SIP system will allow the designer to create a building using a minimum of custom-sized wall elements. These dimensions might vary depending on panel type; dry-process panels will likely conform to standard sheet material dimensions (4×8, 4×9), while wet-process panels may conform to the dimensions of the available straw bales. Either way, planning to a basic module size is ideal.

Specification of all connections, intersections and layers requires understanding of the specific S-SIP system. A designer should consider all the points given on the following page.

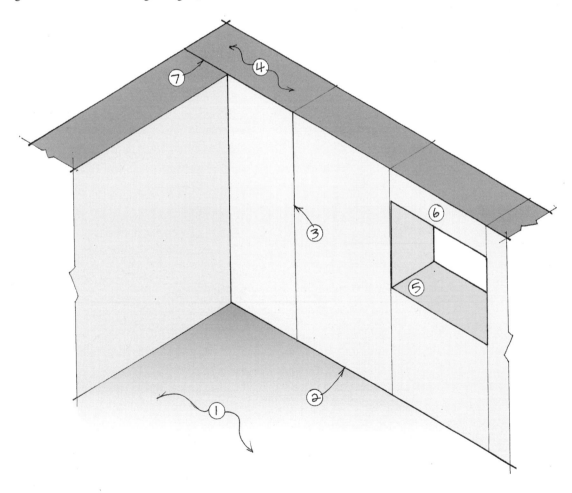

1. **Foundation**
   - Panels are adaptable to any building foundation system. Connection types will be specific to foundation materials and design.
   - Provide foundation floor drains to ensure minor flooding does not overwhelm walls.
   - Overall wall width may require blocking in wood frame foundations under the inner face of the wall.
   - It may be possible to cantilever the S-SIPs beyond the edge of the foundation.
   - Level foundation is important, as walls cannot be adjusted to accommodate unevenness.

2. **Foundation/wall connection**
   - Specify a suitable moisture break layer between foundation and wall.
   - Specify a suitable air barrier to prevent leakage between panel and foundation. Caulking, adhesive, expanding tape and/or surface tape should be chosen to suit site and code conditions.
   - Provide a toe-up suitable to meet fastening and trim needs and to elevate walls above potential interior floor height.
   - Provide an adequate flashing at the exterior wall base to keep water away from wall/floor connection.

3. **Panel/panel connection**
   - Specify appropriate caulk, adhesive, expanding tape and/or surface tape to prevent air leakage.
   - Panels may not require fastening to each other for structural purposes. If fastening in the field is required, ensure fastening system works with chosen wall sheathing/finish.
   - Panels may be connected at the top by framing lumber or sheet material that spans joints.
   - Plastered walls may require reinforcement for finish coat at seams.
   - Additional sheathing/cladding should span over panel joints.

4. **Panel-to-roof connection**
   - Specify placement of roof bearing point(s); can be outside edge of wall, inside edge of wall or center of wall depending on design of load path.
   - Ensure that panel top plate includes appropriate fastening surface for roof members.
   - Ensure that panel top plate includes appropriate blocking across the width for roof members.
   - Framing lumber top plate can be added to fasten and align panels and to provide for roof attachment.
   - Design for appropriate fire blocking at top of wall.

5. **Window details**
   - Placement of window in the depth of the wall will determine sill requirements and return/trim details.
   - Provide positive drainage and proper drip edge on exterior sill.
   - Provide adequate flashing at top edge of window.
   - Rounded, flared or angled window openings will require appropriate framing details.

6. **Header details**
   - Headers to be designed according to code requirements.
   - Bearing point may be outside of wall, inside of wall or center depending on design of load path.

7. **Wall corners**
   - Ensure adequate air sealing at corners.
   - If required, ensure proper framing placement for connection in the field.
   - Plastered walls may require reinforcement for finish coat at interior and exterior corners.
   - Details for additional sheathing/cladding should be provided for interior and exterior corners.

Chapter 8

Construction Procedure

H AVING DECIDED ON ALL OF THE DESIGN parameters for the panels, the construction process is straightforward, and is quite similar for all panel types. The drawings of the construction procedure outlined here use "generic" framing components. The assembly of the elements selected in the previous chapter is common to all framing types.

S-SIPs are best constructed with the wall lying horizontally on a flat floor or raised workstation. By keeping all work at ground level, labor input is reduced and worker safety is enhanced. With a level floor or platform under the wall, achieving a square and plumb wall is straightforward.

Individual frame components are assembled to create the box.

1. **Assemble framing elements using the specified fastening system.** Apply caulking or adhesives to each seam as required, and ensure the frame is square.

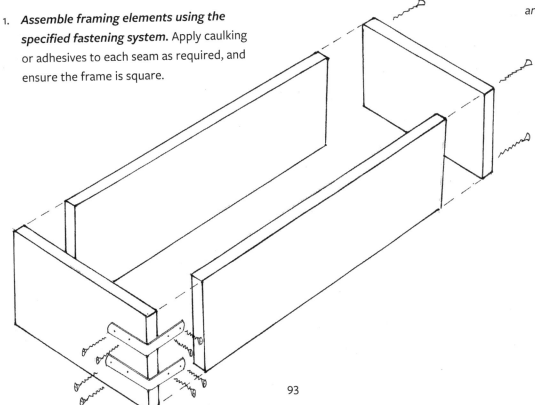

2. **Attach any internal bracing for the bottom face of the panel.** The frame should be checked for square and plumb.

At this point, dry-process and wet-process panels differ in the construction steps.

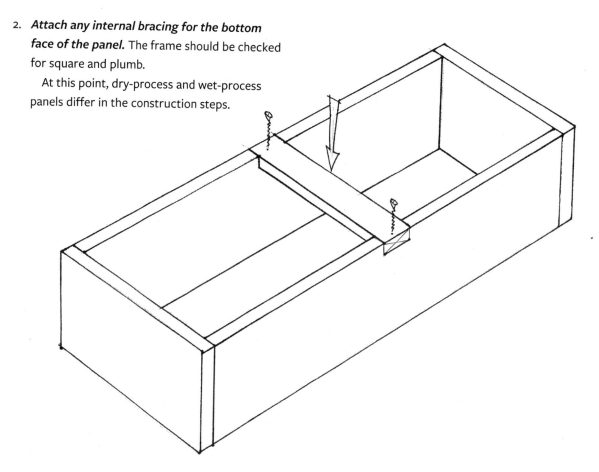

A cross brace is installed on this frame to keep the sides restrained and provide attachment for the dry sheathing.

Wet Process

3. ***Fasten a temporary backing material to the frame.*** This will act as a form for the plaster. Ensure the fasteners are easy to remove later.

    The plaster will take on the texture of the backing material. If a final finish coat is to be applied on site, this texture can be intentionally rough to provide mechanical grip for the finish.

    Formwork to create intentional patterns or forms in the plaster can be inserted into the form.

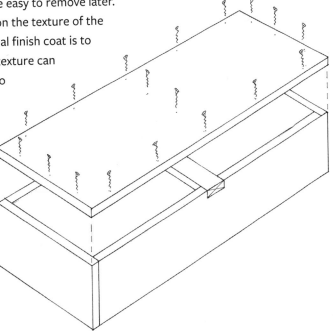

Plywood is used to make a removable form for this wet-process panel. The form can accompany the wall to the site and be removed after installation. CREDIT: KATIE HOWARD

4. ***Flip frame.*** It may be beneficial to temporarily pin or fasten the sides to keep the frame square. Attach any mesh required to the lower sides of the frame and bracing.

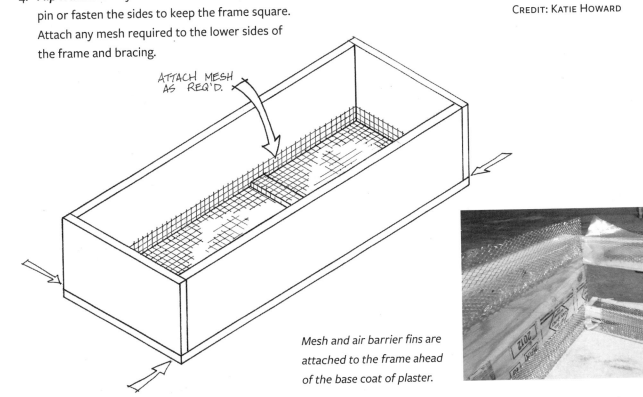

ATTACH MESH AS REQ'D.

Mesh and air barrier fins are attached to the frame ahead of the base coat of plaster.

5. ***Place plaster into frame to the appropriate depth.*** Plaster volume can be calculated and/or a simple depth gauge can be used. Achieve a smooth, consistent layer of plaster, being sure to pack plaster into corners and to fully surround any mesh. The plaster should be free of voids, and mechanical vibration may be required to get rid of air pockets.

At this stage, if any electrical conduit, wiring and/or boxes are intended to be installed on the lower face of the wall, they would be placed now.

Plaster is placed into frame, leveled and then vibrated to eliminate voids.

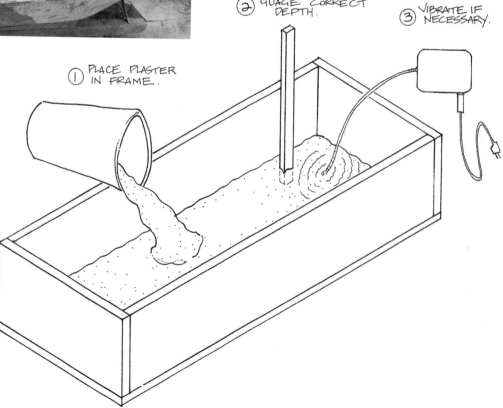

① PLACE PLASTER IN FRAME.

② GUAGE CORRECT DEPTH.

③ VIBRATE IF NECESSARY.

6. **Apply a runny plaster mix to one side of each bale.** The plaster must penetrate into the straw as deeply as possible to create a strong bond, as the wet plaster in the frame will not make a strong connection to the bare straw in the bale. The plaster in the bale and the plaster in the form will create the required unified coat of plaster. The more completely the surface of the bale is coated in plaster, the less likelihood of cracking when the wall is in service.

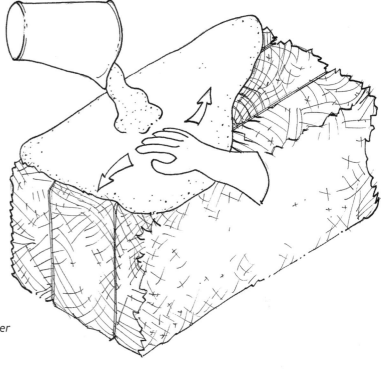

Runny plaster must be worked vigorously into the surface of the bale to ensure a good bond between plaster and straw.
Credit: dan earle

7. **Insert bales into frame with the plastered side down.** As each course is installed, knock the bales tightly against one another. Stuff any gaps or voids with straw. Some builders cut the strings to encourage the bales to expand horizontally and fill voids.

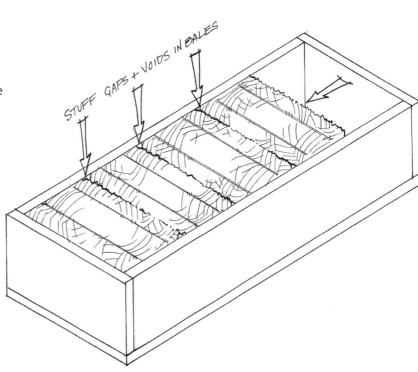

Rolls of straw/clay can be used to stuff gaps and voids along the edges of panels and between bales.

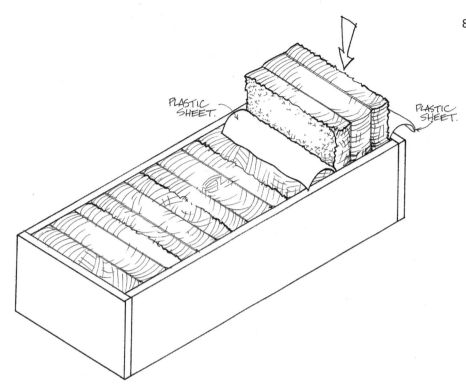

8. ***Place the final bale or course of bales.*** A plastic sheet over the second-to-last course of bales and the top plate will reduce friction and help the final bale slide into place. The fit should be very tight. Once the final course is in place, withdraw the plastic sheets. Electrical conduit and/or boxes may be installed at this point if required.

9. ***Pour plaster on top side of panel.*** Begin with a thinner mix that is scrubbed deeply into the straw, coating the bales completely and ensuring strong contact with the straw. Then pour remaining thickness of plaster. Depending on the frame type, the frame may act as a depth gauge, or temporary forms around the sides of the panel will be the forms.

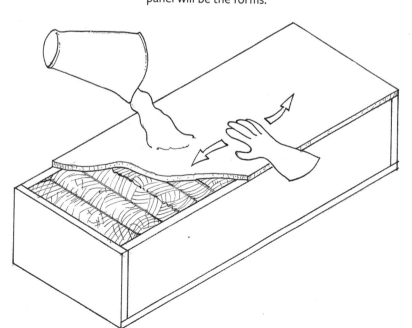

Plaster is first worked into straw surface by hand to ensure a strong bond.

10. ***Screed and/or trowel the plaster to the desired degree of finish.*** This may be the final finish for the wall, or a base for a final site-applied coat and can be treated accordingly.

The plaster on both sides of the panel will need to cure or dry before being handled. This can take 1–14 days, depending on the type of plaster used. Moving the panel before the plaster is fully cured can result in cracking and compromising the working strength of the panel.

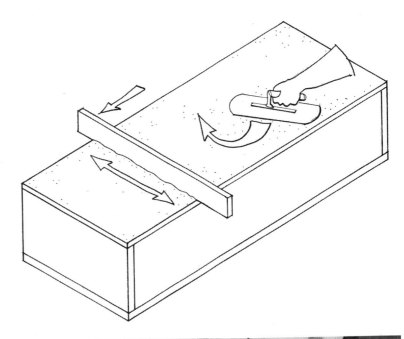

Note: Some builders of wet-process panels do not apply plaster to the wall until the panels have been installed on site. In this style of S-SIP, the bales are installed into the frame and delivered to the site with the straw exposed. All the stages for preparing the wall and applying plaster are performed on site, not in the production facility.

This approach misses out on some of the key advantages of prefabrication by subjecting the plaster to unpredictable site conditions and requiring the application of successive coats of plaster, rather than the one-coat system used if the plaster is applied at the production facility. The advantage is that the panels are much lighter to transport and handle.

These S-SIPs were delivered to the building site unplastered, and all the required coats of plaster are applied in situ.

Screeding will level the plaster. The surface can be troweled if a smoother finish is desired.

Dry Process

3. **Fasten sheathing onto or into frame.** Use specified fasteners and caulking/adhesives as required.

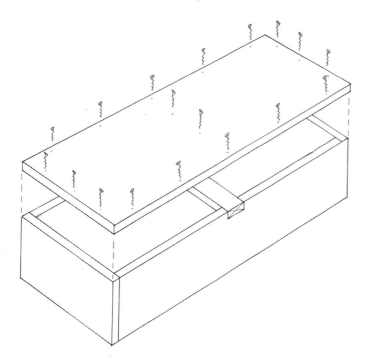

The adhesive provides air sealing around the panel and structural strength to the frame and sheathing.

4. **Flip frame.** It may be beneficial to temporarily pin or fasten the sides to keep the frame square.

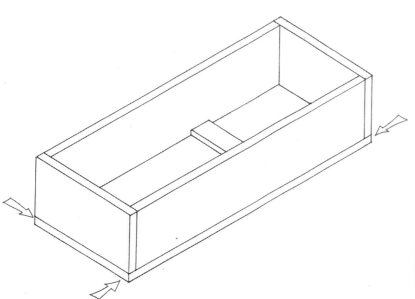

Dry panel (with wood fiber board sheathing) is flipped and ready to receive bales.

As each course of bales is added, they are pounded tightly into one another. CREDIT: BEN BOWMAN

The last bale fits very tightly into the frame. A plastic sheet reduces friction and helps the bale be persuaded into place. CREDIT: BEN BOWMAN

5. **Insert bales into frame.** As each course is installed, knock the bales tightly against one another. Stuff any gaps or voids with straw. Some builders cut the strings to encourage the bales to expand horizontally to fill voids.

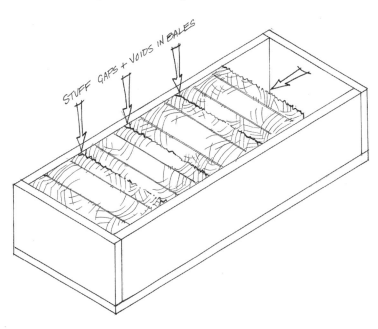

6. **Place the final bale or course of bales.** A plastic sheet over the second-to-last course of bales and the top plate will help the final bale slide into place. The fit should be tight. Once the final course is in place, withdraw the plastic sheets. Electrical conduit and/or boxes may be installed at this point if required.

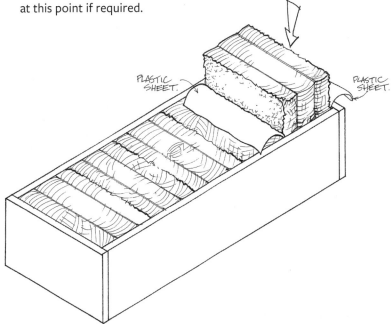

7. ***Attach sheathing onto or into frame.*** Ensure there are no voids or gaps in the insulation. Use specified fasteners and any caulking or adhesive required.

   The panel is ready for moving when any caulking/adhesives used are fully cured.

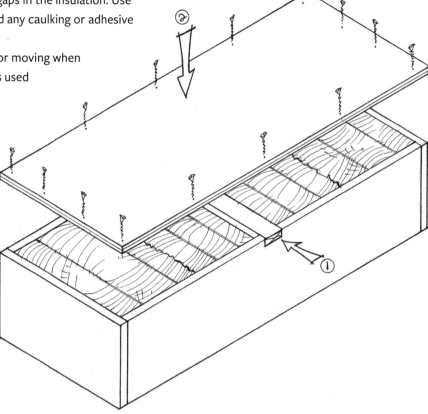

Above: *The dry sheathing is nailed or screwed as required to the top of the frame, enclosing it.*

Right: *Dry panels can be moved and stacked immediately after construction.*

Production Drawings and Process

Detailed shop drawings are important for the S-SIP production process. A complete shop drawing should include:

- Wall numbering system.
- Overall dimensions of each wall.
- Dimensions for each top, side and bottom plate.
- Placement information for conduit and electrical boxes, if required.
- Placement information for any special wall openings (HVAC, electrical feed, plumbing).
- Placement information for any additional framing requirements (ledgers, additional structural elements, attachment points for cabinetry).

It is a good idea to have a printout sheet for each wall panel that contains all the above information and remains with the wall panel throughout construction, shipping and installation. If different people handle elements of the panel construction and/or handling processes, having them sign off on the sheet is a good way to ensure quality control.

CREDIT: BEN BOWMAN

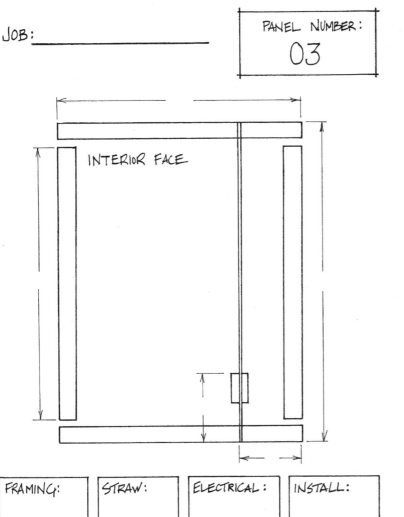

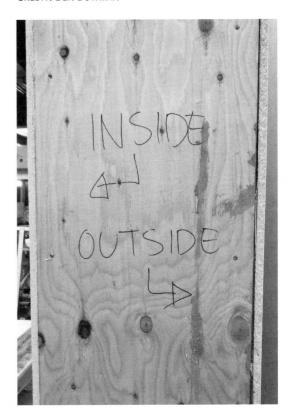

In-house labeling and numbering of panels helps ensure they are built correctly.

Jigs

To produce S-SIPs in quantity, it is worth considering building or commissioning jigs to help keep panels square and properly sized. Jig tables can also place work at a comfortable height for builders.

The specific details of any jig will be particular to the type of panel being produced. In general, the jig should offer flexibility and adjustability to allow builders to quickly set location, working height and panel dimensions, and then be able to reliably fix these dimensions while building a panel.

The adjustable sides of the jig can also be used to compress the straw bales in the frame if this is required.

A forklift may be a necessity for wet-process panels, or large-sized panels of any type.

Handling and Transportation

Panels built off site will require handling within the construction facility and transportation to the job site. Dry-process S-SIPs can weigh 10–15 pounds per square foot ($49–73 \text{ kg/m}^2$) and wet-process S-SIPs can be in the range of 30 pounds per square foot (146 kg/m^2), making them difficult or impossible to move manually.

S-SIP facilities often employ a forklift to move panels within the facility and to load them onto trucks. Pallets or spacers are used to ensure the forklift can be configured to allow the forks to lift the panel from the side or the top/bottom. This choice will be based on typical panel size and the arrangement of panels on the transporter.

Panels can also be moved without the use of a forklift. A hand truck or pallet jack can be used to provide movement around the facility and prepare the walls for loading.

Some facilities have used a purpose-built A-frame that can both move walls and lift them from horizontal to vertical position.

Lifting points

For S-SIPs fabricated off site, the builder will have to decide how lifting and placement of the walls will be handled.

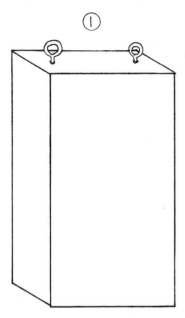

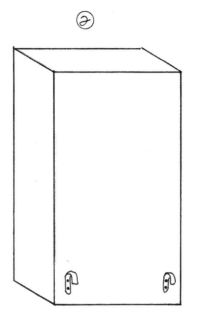

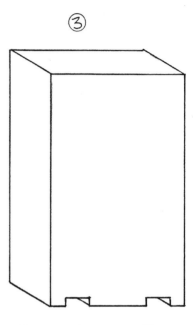

1. ***Top mounted lifting points.***
   Temporary lifting hooks are screwed into the frames at the top of the wall, or attached along the sides of the wall.

 Advantages:
- Top lifting is the easiest system to attach straps and maneuver walls.
- Walls can be set directly on foundation.
- Walls shipped lying horizontal are easy to stand up.
- Many inexpensive hook systems available.
- System best suited for panels with structural sheathing.
- Two hooks may be sufficient.
- Hooks can be reused.

 Disadvantages:
- Applies tensile stress on all frame joints and any plaster skins. Design must account for tensile loading.
- Depending on frame style, four hooks may be required.
- Holes left by hooks may need filling.

2. ***Bottom-mounted lifting points.***
   Temporary lifting hooks are screwed into the frames at the bottom corners of the wall.

 Advantages:
- Walls can be set directly on foundation, no interference with bottom plate.
- Secure anchor points, wall cannot slip on straps.

 Disadvantages:
- Hook fasteners leave holes in face of wall, not suitable for walls delivered with finished surface treatment.
- Four hooks required.
- Hooks protrude from face of wall, can prevent stacking or tight tolerances during transportation.

3. ***Strap notches on panel base.***
   A notch is created on the bottom plate to allow lifting straps to be passed under the wall and withdrawn after wall is placed.

 Advantages:
- No need for expensive hooks.
- Least stress on frame and plaster.

 Disadvantages:
- Extra labor required to build bottom plate with notches.
- Extra labor required to fill notches on site, extra care needed for air tightness.
- May not be compatible with all foundation systems for removal of straps.
- Difficult to lift walls from horizontal position due to slippage inside the straps.

Chapter 9

Installation

S-SIP INSTALLATION is relatively straightforward. Adaptions for specific styles of panel and foundation materials will need to be made, but the process follows the same steps:

1. *Make a site inspection if building walls off site.*

   • If using a crane or other lifting equipment, inspect the site to ensure that the delivery truck has adequate access and that the lifting equipment can be placed at an appropriate distance from the delivery vehicle and the building. Check for overhead wires and proximity of nearby buildings. A crane will require a relatively flat place for staging and be able to swing the boom from delivery vehicle to foundation. Lift trucks will require enough room to maneuver walls around the site and approach each side of the foundation.

Power lines may be present at the installation site. A tall crane may be able to lift panels over wires and other obstacles.

Lift trucks can be used to install S-SIPs on a site where there is adequate space to approach the foundation from all sides.

2. ***Prepare the site for installation.***

- Mark the foundation with the position of each panel. Check measurements to ensure that the panel sizes and the foundation size match, and create a clear marking system that corresponds to the numbering of the panels. The marking system should be legible from a distance, and be waterproof.

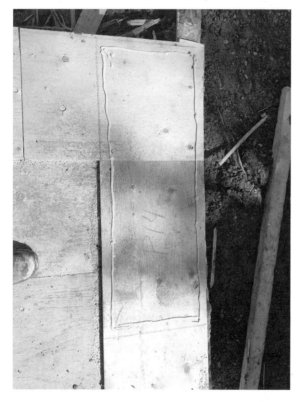

The position and number of each panel has been marked on the foundation. Caulking has been laid to seal the panel to the foundation.

- Prepare any foundation elements required to fasten the walls in place. This can include sill plates or other attachment points.

3. ***Choose the position for the first panel to be installed.***

- It is best to start the installation on the opposite side of the building from the position of the lifting equipment so that walls do not need to be lifted over top of those already installed.
- Add any required sealant, tape and/or moisture break required between panel and foundation.

4. ***Put the first wall into position.***

- Using the foundation markings to set the wall in place. Minor adjustments can be made using a wedge and lever or ratchet straps.

It is best to begin placing walls at the furthest point from the crane.
CREDIT: KATIE HOWARD

5. ***Anchor the first panel and attach temporary bracing.***

   - Bolt, screw or otherwise anchor the panel as required. Temporary bracing should be fastened to keep the wall plumb and to prevent wind or contact from toppling the panel. This bracing should remain in place at least until all walls and corners are joined, and will ideally remain in place until a roof or floor system is in place.

6. ***Apply sealant, expanding tape or other air sealing material between panels.***

   - Ensure that the intended seal between wall sections is achieved. Use a ratchet strap or wedge and lever to draw panels tightly together. Creating a system of checks for the installation crew can help to ensure that a seal is reliably made.

Once the panel is in position, the team ensures the wall is plumb and then attaches temporary bracing to keep it secure.

Toe screwing the base plate with structural screws can temporarily secure the panels. This may be the chosen attachment method.

7. ***Complete the installation and check walls for plumb and square. Complete fastening of panels to foundation and to each other.***

- Ensure that the completed panels create square corners, that the walls are plumb and that the tops of the walls align. Adjustments can be made using wedge and lever, ratchet strap and temporary bracing.
- Once the walls are in position, complete the fastening of panels to one another as per the original design specifications. In some cases, this may involve an additional top plate to connect and fasten the panels.

8. ***Caulk or tape face of joints, as required***

- If any caulking or tape is required on the faces of the joints, including the top joint, this can be done once the walls are completely fastened, plumb, square and attached together.

For this building, seams between panels, framing sections and floors are taped to provide air sealing.

The first corner is erected and bracing is installed to keep panels square and plumb.
CREDIT: IAN WEIR

Additional framing lumber can be used to connect panels together.
CREDIT:
JOHN GLASSFORD

Chapter 10

Tilt-up Construction

MUCH OF THE FOCUS IN THIS BOOK has been on S-SIPs that are fabricated in an off-site facility. However, panels can also be constructed on site, positioned on the foundation to allow for a tilt-up installation.

This process is well suited for owner-builders who want to take advantage of the speed and simplicity of building in panelized form, but do not need to worry about the longer site times required to do the building *in situ*. Small contractors undertaking one or two builds per season may also benefit from this approach, as the costs of operating a production facility may offset any on-site time savings.

There are many advantages to making on site S-SIPs over site building with straw bales (or other wall systems). Tilt-up panels built on site can remove some of the pressure of unpredictable weather, as panels can be built individually or in small batches that can be completed in one day. At the end of the day, the panels are either weather ready or relatively easy to cover.

The speed of constructing walls lying horizontally is much greater than walls built vertically. Framing crews have been assembling walls and attaching sheathing horizontally for decades because of the time savings, and those

using natural materials like straw bale can benefit from following their lead.

For an owner-builder, tilt-up S-SIPs can be made without the need for a larger crew, as is common for site-built straw bale walls. One or two people can build the frames, install the bales and apply plaster and/or sheathing in a horizontal position much more easily than having to lift and place framing, bales and other materials using ladders and scaffolding.

All of the wall elements and construction processes outlined in this book apply to on-site construction with little or no modification.

Tilting panels into place may be possible to do manually, especially for dry-process S-SIPs. Heavier wet-process panels may require some mechanical assistance, or panels can be made small enough that the on-site crew is capable of tilting them. Hydraulic jacks, gin poles, winches, bucket trucks or loaders could all be used to tilt heavy panels into place.

A crew building S-SIPs on site could also choose to build the load-bearing walls of the building, and then build the roof onto these walls before continuing to build the remaining walls under cover of the roof.

Tilt up S-SIPs are ideal for building on site and in place. Using the same techniques as off-site fabrication, a hinged panel is built, plastered and lifted into place by Sven Johnston and crew at Sol Design. CREDIT: SVEN JOHNSTON

Chapter 11

Finishing Installed Walls

A WIDE RANGE OF FINISHES can be applied to S-SIPs. The type of panel (wet-process or dry-process), climate conditions, code requirements and aesthetics will have guided decisions regarding finishes for a particular installation. All necessary provisions for finishes will have been taken into account during the construction of the panels.

Plaster Finishes

Wet-process panels

A skim coat of finish plaster can be added directly to S-SIPs that are built with the base coat of plaster as the delivered surface.

1. ***Tape and mud all panel joints.*** This step is required to help prevent cracks from occurring at joints, and the process is the same as that used in covering conventional dry wall seams. A joint compound is used to fill the crack between panels, and then a reinforcing tape (commonly a fiberglass mesh tape) is embedded into the wet joint compound. Additional joint compound is then added over the tape and feathered out onto the surface of the panels to create a smooth, well-blended seam. This process may require more than one coat of joint compound and may require sanding between coats, depending on the thickness of the final plaster.

   The joint compound used must be appropriate for the location, the weather exposure and the type of plaster being used for the finish coat. Typically, the joint compound is made from the same binder material as the plaster.

All wooden surfaces to be covered with plaster have been prepared with a mesh tape to provide mechanical grip for the plaster.

2. ***Prepare, mesh and apply base plaster to cover any exposed wooden elements.*** Wood being covered with plaster is typically covered to prevent the wood from absorbing water from the plaster, swelling, and then causing a crack when it dries and shrinks. Building paper can be stapled over wood, or a surface preparation (some form of sand/glue) can be brushed or sprayed onto the wood. The surface can then be meshed and prepared with joint compound.

3. ***Apply any required trim.*** Some trim options will dictate that the trim be installed prior to the finish coat, with the plaster meeting the edge of the trim.

4. ***Prepare entire wall surface, if necessary.*** With the joints and wood elements covered, the plaster may require a bonding agent or other form of surface preparation on the full wall surface, especially if the base coat is smooth and does not offer much mechanical key for the finish coat to grip. Taping at all edges where plaster meets other building elements will help ensure a clean finished result.

A continuous coat of finish plaster can be applied as the final surface of S-SIPs after joints have been prepared.

CREDIT: JEN FEIGIN

5. ***Apply finish plaster.*** The final coat of plaster will vary in thickness depending on the type of plaster and the desired finish. For exterior plaster, there may be code requirements that should be consulted regarding plaster type and mix ratios. The use of plaster stops will help to gauge thickness.

Be sure that plasters are applied in the right weather conditions and that all required curing/drying processes are followed for best results.

6. ***Apply any surface treatments to cured plaster.*** There may be a requirement or desire for a final surface treatment on the finish plaster, to provide color, texture and/or protection. On the exterior, silicate dispersion paints can provide valuable water protection without compromising permeability. On the interior, a wide range of paints can be applied to meet needs.

Dry-process panels

A skim coat of finish plaster can also be applied to panels that do not have a base coat of plaster. The steps used will depend on the type of substrate being plastered, but follow a common progression.

1. ***Seal and prepare panel seams.*** If plaster will be applied directly to wall substrate, seams will need to be prepared with joint compound or plastering tape.
2. ***Apply appropriate surface preparation.*** A bonding agent may need to be applied to the wall substrate to allow plaster to adhere properly. This can be in the form of a spreadable material, a plaster mesh, or a plaster breather membrane.
3. ***Apply any required trim.***
4. ***Apply finish plaster.***
5. ***Apply any surface treatments to cured plaster.***

Direct-applied Finishes

A variety of sheathing materials can be directly applied to the surface of either wet- or dry-process S-SIPs. Permeability, durability and drainage must all be considered when selecting a direct-applied sheathing. In general, this type of finish is applied to the interior face of the walls, but may be appropriate for some exterior applications. In general, it is best to use *ventilated rain screen finishes* on the exterior.

1. ***Weather and air barrier.*** Ensure appropriate air and weather sealing at all joints. This may involve caulking/taping seams and/or applying a sheet-style membrane on the exterior of the building. Be sure to check local codes, which may mandate an exterior weather membrane.
2. ***Apply finishing material.*** Use specified fasteners and/or an adhesive recommended for the type of finishing material. Attachment points will vary depending on the construction style of the frame, the type of sheathing

used on the panel, the location of bracing in the frame and the type of finish being used. For interior drywall application, a mix of screws and joint compound applied to the wall surface can provide connection to the walls. Joints are then taped and mudded as per typical drywall installation.

3. ***Apply trim and surface treatments.***

Strapped or Vented Finishes

Any type of siding or wall finish can be applied over strapping/furring on S-SIPs.

Interior finishes

Strapping can be applied to the wall surface to meet spacing requirements for the finish material and to create a desired amount of space between wall surface and finish.

1. ***Apply strapping to the wall.*** Use available fastening points in the wall framing and the sheathing to attach strapping material. In general, strapping is applied vertically to allow for solid fastening points into the frame at the top and bottom of the wall, but horizontal strapping is also possible if required. If spans between attachment points are far apart, an adhesive may also be used.

2. ***Apply finishing material to the strapping.*** The finishing material may require additional steps for dealing with joints. Follow appropriate manufacturer's instructions and code requirements.

3. ***Apply trim and surface treatments.***

Exterior finishes

Creating a ventilated rain screen will provide the wall with a resilient finish, keeping precipitation from coming into contact with the wall, and providing a ventilated plane for drying the wall and the siding.

1. ***Weather and air barrier.*** Ensure appropriate air and weather sealing at all joints. This may involve caulking/taping seams and/or applying a sheet-style membrane on the exterior of the building. Be sure to check local codes, which may mandate an exterior weather membrane.

2. ***Apply strapping to the wall.*** Use available fastening points in the wall framing and the sheathing to attach strapping material. In general, strapping is applied vertically to allow for solid fastening points into the frame at the top and bottom of the wall, but horizontal strapping is also possible, if required. If spans between attachment points are far apart, an adhesive may also be used.

3. ***Protect ventilation channel.*** The ventilation channel behind the siding requires a free flow of air, but should be sealed against insects and rodents. Breather strips can be made or purchased to completely fill the spaces between wall, strapping and siding in a manner that will provide ventilation and protection.

Breather strips are cut to fill the areas between strapping, allowing air circulation but preventing insects from intruding. Strapping at 45 degrees provides structural bracing as well as ventilation channels.

Once the S-SIPs have been strapped, siding installation follows conventional application techniques and manufacturer instructions.
CREDIT: KATIE HOWARD

At the top of the wall, the ventilation channel can open into the soffit or be vented to the exterior. Soffit venting is acceptable as long as there is adequate ventilation in the soffit and in the roof to deal with the additional moisture load. Face venting requires a good seal at the top of the channel and shielding to protect against water and pest entry.

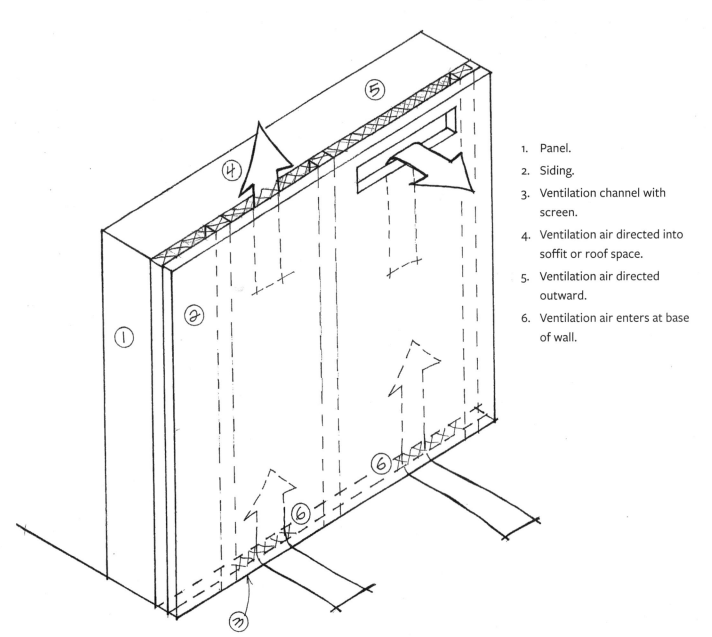

1. Panel.
2. Siding.
3. Ventilation channel with screen.
4. Ventilation air directed into soffit or roof space.
5. Ventilation air directed outward.
6. Ventilation air enters at base of wall.

Above: Designed by Harvest Homes, built in 2003 near Erin, Ontario.

Home Alive!, The House That Thinks, Drinks and Breathes™ was first constructed on the floor of the National Home Show in Toronto, Ontario before being reassembled on its current site. This portability can be a real advantage of S-SIP construction. CREDIT: BEN POLLEY/HARVEST HOMES

Above and bottom right: Designed by fuentesdesign architects, built in 2012 near Boulder, Colorado.

The Marshall Mesa home was designed to meet the rigorous Passive House standards for energy efficiency. CREDIT: DANE CRONIN

Above: Designed by Jen Feigin and Chris Magwood, built in 2012.

This urban S-SIP home was built on an infill lot in Peterborough, Ontario, and reduced energy needs by 75% over a code-built home. The mix of wood siding and cedar shingles allow the house to blend into the existing neighborhood.

CREDIT: CHRIS MAGWOOD

Right: *A finish plaster of red clay and a bright white lime paint adorn the S-SIP walls, adding warmth and texture to the panel surfaces.*

CREDIT: DANIEL EARLE

Above: Designed by Jen Feigin and Chris Magwood and Our Cool Blue Architects, built in 2009 in Peterborough, Ontario.

The Camp Kawartha Environment Centre was the first use of prefab straw bale walls for a public assembly building in Canada. The Centre provides education about sustainable building and living options for youth and adults. CREDIT: CHRIS MAGWOOD

Below: *In the main meeting room of the busy Camp Kawartha Environment Centre, built-in benches in the window areas between the panels offer additional seating as well as storage. Lighting is incorporated into the headers over the windows. Both these design elements make good use of the 16-inch wall width.* CREDIT: CHRIS MAGWOOD

A timber frame wrapped with S-SIPs keeps the home of Jane Strong in Sharon, Connecticut warm when the snow is blowing. CREDIT: JANE STRONG

Traditional timber framing can use S-SIPs to replace foam-based SIPs and keep the overall ecological footprint of the building low. CREDIT: JANE STRONG

DanEarle

Above: Designed by Jen Feigin and Chris Magwood, built in 2010 in Peterborough, Ontario.

Prefabricating the wall panels for this Habitat for Humanity house allowed the project budget and timeline to match those of more conventional builds by the not-for-profit housing organization. A rainscreen of horizontal siding helps the house fit with others on the street. CREDIT: DANIEL EARLE

Below right: *Despite the "normal" appearance of the walls in this Habitat for Humanity, a small "truth window" reveals the straw bales inside the S-SIP walls.* CREDIT: DANIEL EARLE

Above: Designed by HavenCraft Design, built in 2013 near Millbrook, Ontario.

This house is the hub of the busy Circle Organic Farm. Painted wooden rainscreen siding protects the walls from the driving rains found at the top of this hill site. CREDIT: CHRIS MAGWOOD

Left: Designed by HavenCraft Design, built in 2013 near Millbrook, Ontario.

Clay finish plaster joins the S-SIPs and the framed section at the bay window seamlessly. Angled window returns are a common feature of older farmhouses in the area, and were easily built into this home. CREDIT: JEN FEIGIN

Top: Designed by Architekten für Nachhaltiges Bauen GmbH

The North German Centre for Sustainable Building in Verden was completed in 2014. At 2,223 m² and five stories, this "ecological lighthouse" is one of the largest S-SIP buildings in the world, and it met the stringent Passive House energy efficient standards. CREDIT: DIRK SHARMER

Center left: Designed by Nicolas Koff and built by Evolve Builders Group in 2015 in Ancaster, Ontario.

Panelized construction allows for the clean modern lines on this award-winning residence, while still delivering great energy efficiency and low environmental impacts. CREDIT: NICOLAS KOFF

Center right: S-SIPs can provide sharp, square edges and open up a range of design possibilities that are difficult to achieve with site-built straw bale walls, as seen in this bright, open design. CREDIT: NICOLAS KOFF

The design of The Gateway Building uses the form of the long, tall S-SIPs to inform the overall design.
CREDIT: MAKE ARCHITECTS

Center left: *The Gateway Building at the University of Nottingham in England is a 3,100 m² education and research facility designed by Make Architects. The design was influenced by the campus's agricultural heritage and strong sustainability policy, and uses panels built from bales harvested on the university's own farmland adjacent to the building site.*
CREDIT: MAKE ARCHITECTS

Center right: *The Red Rock Marina Interpretive Centre in Red Rock, Ontario, features an educational exhibit hall and multipurpose rooms for art exhibits, workshops and conferences. The S-SIP panels are wrapped around a timber frame, and the*

trim pattern displays one straightforward way of working with panel seams.
CREDIT:
GERALD SARRASIN

Below: *The offices and warehouse of Solacity, a distributor and retailer of renewable energy products in Kemptville, Ontario, shows how S-SIPs can be effectively used to create large, simple buildings with high levels of energy efficiency at very reasonable costs.*
CREDIT: CHRIS MAGWOOD

Above and center right: *This home in Liptovská Kokava in northern Slovakia is by Createrra. It has Ecococon walls and features a wooden rainscreen on the exterior and clay plaster and a masonry heater on the interior.*
CREDIT: MILAN HUTERA/ARCHIV CREATERRA

Below right: *Modern form blends with traditional materials in this prefab bale home in Bad Deutsch-Altenburg in Lower Austria using the Ecococon wall system.*
CREDIT: MILAN HUTERA/ARCHIV CREATERRA

Above and center left: *Wood and plaster define traditional and modern style in Predajná, central Slovakia.*
CREDIT: MILAN HUTERA/ARCHIV CREATERRA

Below left: *The Ecococon wall system is brought to site unplastered, and the sheathing or plastering is done in situ.*
CREDIT: BJORN KIERULF/ARCHIV CREATERRA

Above: *The Hayesfield Girls'
School in Bath, England, uses
both S-SIP walls and a straw
bale roof cassette system
from ModCell to create an
excellent thermal package for
its nucleus building of labs
and classrooms, built in 2013.
It won a "Gold" Green Apple
Award 2013 from The Built
Environment and Architectural
Heritage category.*
CREDIT: MODCELL

Center and below right: *Using ModCell prefabricated straw bale
walls, Inspire Bradford built a unique 30,000 ft² business park that
combines 14 service offices and 14 workspaces with a community
facility in two buildings.* CREDIT: MODCELL

Above: *The LILAC (Low Impact Living Affordable Community) project is a member-led, not-for-profit cooperative society. They built a community of 20 beautiful homes in Bramley, West Leeds on an old school site using ModCell panels. Their community will include a mix of one- and two-bed flats and three- and four-bed houses.* CREDIT: MODCELL

Center left: *The seven homes in St. Bernard's Road in Bristol, England, were the world's first straw bale homes to be commercially available on the open market. Built by Sustainable Britain, a division of Connolly & Callaghan, they used ModCell panels and a brick rainscreen cladding to match local aesthetics.* CREDIT: MODCELL

Below left: *The Knowle West Media Centre provides opportunities for participation in media arts by young people living in one of the most disadvantaged areas of Bristol, England. Built in 2008, it was the winner of the South West C+ Carbon Positive award.* CREDIT: MODCELL

Above: *S-SIP panels can travel upright or lying flat on a truck bed. Enough panels for one average-sized house (2,000-2,500 ft²) can fit on a typical truck.* CREDIT: CHRIS MAGWOOD

Center left: *Prefabricated walls are easily moved to upper stories with a crane.* CREDIT: CHRIS MAGWOOD

Center right: *Panels are placed with a small team of builders guiding each panel into place. These panels feature built-in air control flanges that allow for a tight air seal between panels and connecting framing.* CREDIT: CHRIS MAGWOOD

Below: *The third SITUPS panel home by Huff'n'Puff Construction, for a client in the Mudgee District of New South Wales, is ready for plastering.* CREDIT: JOHN GLASSFORD/ HUFF'N'PUFF CONSTRUCTION

Chapter 12

Maintenance and Renovations

S-SIP WALLS DO NOT REQUIRE any regularly scheduled maintenance regimen. The surface finish, siding and sheathing protect the wall core. As long as these surfaces are maintained adequately, the wall core itself will not need inspection or maintenance.

The protective layers on the wall are easily inspected visually. Ensure that paints are in good shape, with no cracking, peeling or wear. Repaint on a schedule recommended by the paint manufacturer. Siding should be structurally sound, with no cracking, warping, delaminating or other visually obvious deterioration. When siding issues arise, repair or replace as needed.

In cases where paint and siding have failed due to lack of maintenance, it is prudent to inspect the wall sheathing to assess its condition. Plaster or sheet materials should be inspected for cracks, water damage and other issues. Only if plaster or sheathing show signs of extreme deterioration will it be necessary to inspect the straw inside the wall.

Maintenance regimens for different paints and sidings can vary, but typically range from 15–50 years. Only complete failure of these layers would prove problematic for the S-SIP behind.

Water Damage

Should the walls experience extreme wetting, first inspect the plaster and/or sheathing to assess the damage. If these layers show water damage or signs of complete saturation, it is wise to measure the moisture content of the straw bales. This can be done by drilling small holes

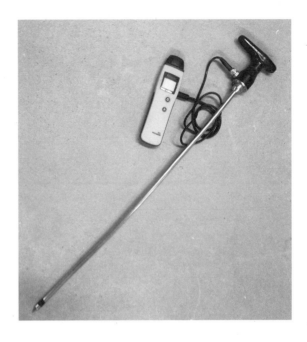

Moisture meter.

through the plaster/sheathing and inserting a bale moisture meter to obtain readings at a variety of locations in the wall and depths within the wall.

These readings can be taken every few days for up to two weeks, and if a strong drying trend is detected, then the walls may dry out without further intervention. This has proved to be the case for many straw bale walls that have experienced flood damage or high-volume roof leaks; this is one of the great advantages of a permeable wall system. The use of fans and/or dehumidifiers can speed up this natural drying process.

If moisture levels in the walls do not decrease at a comfortable pace, the wall can be cut open and the straw can be inspected. If it is showing signs of deterioration, it can be removed and replaced with fresh straw. Most panel designs incorporate a frame with enough structural capacity to allow for the removal of the straw without dramatically lowering the bearing capacity of the wall. If the frame cannot manage structural loads without the straw, the frame can be temporarily braced while the straw is removed and changed. This process can be repeated for as many panels as have been affected by the water damage.

The replacement procedure is the same as the initial construction process, with bales being installed and then covered with whatever plaster and/or sheathing was used in the initial construction.

Extreme water damage is never a good thing, but repair for S-SIPs is no more difficult or costly than other wall systems.

Wall Modifications

Future remodeling of a building with S-SIP walls is relatively straightforward, and follows the same steps as any renovation. Before undertaking any modifications, always consider the structural, moisture, thermal, air control and aesthetic issues raised by the modification.

Modifications to the surface layers of the wall can typically be performed without any need to disturb the wall core. Sheathing and siding layers can be added or removed as long as perforations in the air barrier layer are avoided, or adequately repaired when damaged.

Modifications to the structure of the building must be done in a manner that respects the designed load paths. If posts/sides are removed from bearing walls, proper steps must be taken to redirect loads appropriately. The steps are similar to those taken when dealing with frame walls or post-and-beam walls. New door or window openings can be added to S-SIP walls by removing the siding/sheathing/plaster from the panel, inserting new framing to create the desired opening, and reinsulating the space with straw of the same density and dimension (it is likely that the straw that was removed can be re-used). The straw can once again be sheathed and/or plastered in the same manner as the original panel, and siding and finishes added to match the existing aesthetic.

Joining new addition construction to existing S-SIPs involves similar steps as new wall openings. Ensure that the new abutting wall(s) connect thermally with the straw bale core of the existing S-SIP, and that the air control layer(s) are matched to provide a seamless finish. Siding and surface treatments are handled as required to provide the desired finish aesthetic.

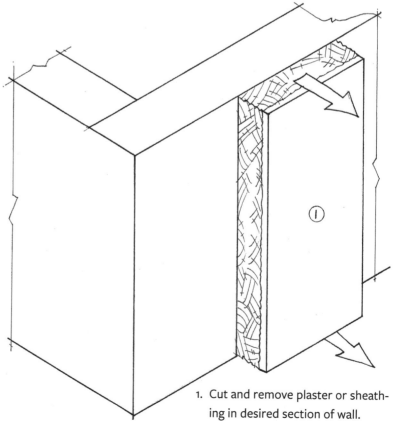

1. Cut and remove plaster or sheathing in desired section of wall.

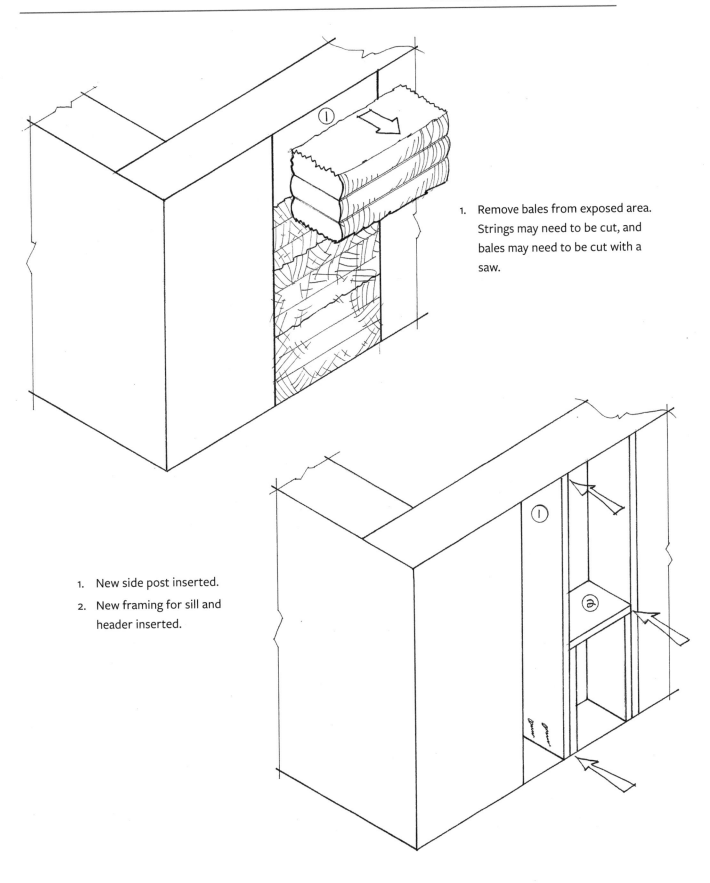

1. Remove bales from exposed area. Strings may need to be cut, and bales may need to be cut with a saw.

1. New side post inserted.
2. New framing for sill and header inserted.

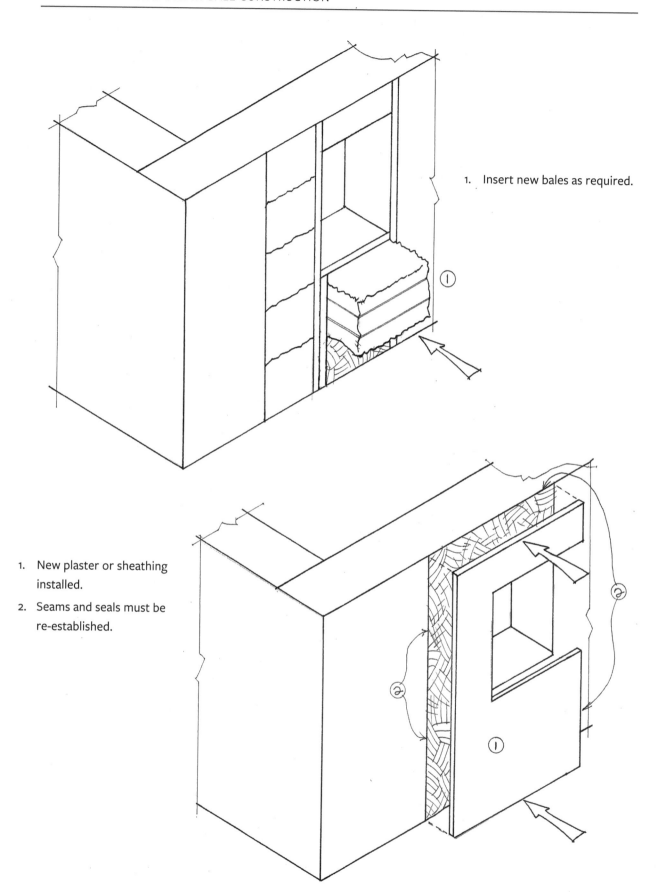

1. Insert new bales as required.

1. New plaster or sheathing installed.
2. Seams and seals must be re-established.

Chapter 13

Building Codes and Permits

T HE REVIVAL OF STRAW BALE CONSTRUC-
TION in North America is now well into
its third decade. The number of straw bale
buildings in the United States is well into the
thousands, according to various regional da-
tabases.[1] A survey of the province of Ontario,
Canada, shows that nearly every municipal jurisdiction in the province has approved at least one
straw bale building. There is plenty of precedent
for being granted a building permit for an S-SIP
structure. In the U.S., the adoption of *Appendix
S — Straw Bale Construction* into the *2015
International Residential Code (IRC)* is a great
asset when applying for a permit. But past precedent and code support for a wall system is not
a guarantee for receiving a permit, as there are
many factors reviewed by building departments
when assessing a permit submission.

Adhering to Local Process

A great number of permit problems arise from
applicants not following the proper permitting
process. Procedural issues far outweigh issues
with a particular building component such as
straw bale walls. Be sure to understand the complete process for obtaining a permit; often there
are multiple stages (planning permission and
site-specific permissions, such as sewer/septic
connections and property setbacks) that must
be completed in a particular order. Most jurisdictions offer guidelines for applying for a building
permit, and these should be followed carefully.
Most building departments have to process a
great number of applications and are typically
understaffed, so incomplete or improper applications do not receive favorable treatment.

Code issues

Both the IRC in the U.S. and the National
Building Code in Canada are model codes,
which are adopted by state, provincial and/or
local authorities in slightly different versions. Be
sure to understand which code to be working
from, and use the most up to date version of that
code document.

Building codes cover a wide range of issues
and topics, and there are many areas in which
a set of building plans may not conform to
the code that have nothing at all to do with a
particular wall construction, like S-SIPs. In the
author's experience as a consultant, a majority of permit denials for S-SIP or other straw
bale-related projects have to do with issues that
are unrelated to the wall system. Plans can have
issues related to planning issues (lot lines and
setbacks, overall height, grading, parking allocation), space allocation (minimum room sizes,
means of egress, staircases and railings, window
size and placement) and services (well/water,
sewer/septic, HVAC) that have nothing to do
with chosen materials or assemblies, and these
are the most common problems. Addressing
them requires an understanding of the codes,
but does not directly influence the use of S-SIPs.

Despite plenty of anecdotal evidence to the
contrary, there is no justification in any North
American building code for denying a permit
because of the use of a straw-based wall system.
At the same time, I haven't heard of any building departments that would answer a general
inquiry about whether or not they will permit a
straw-based wall system with a forthright "yes."
Permits are not granted or denied on the basis of

a single material choice, but for meeting complete sets of requirements that demonstrate the viability and safety of the entire structure.

All codes include provisions for working productively with materials that are not directly recognized by code prescriptions, or materials being used in ways that are not directly prescribed in the code. In making a permit application, you must understand what aspects of the proposed building do and do not meet the prescriptions of the local code. You must also acknowledge that plan reviewers are concerned with specifics, not generalities; permits are granted or denied based on the exact details of the proposal, and if the details are missing, inadequate or in contravention of the code, a permit will not be granted even if the idea is generally feasible.

In areas of the United States that have adopted *Appendix S — Strawbale Construction* in the *2015 International Residential Code (IRC)*, there is definitive code language that can be followed to ensure full compliance with the code (see ecobuildnetwork.org/images/pdf_files/strawbale_code_support/AppendixS_SBConstruction_2015IRC.pdf). It is straightforward to show that an S-SIP designed in accordance with the prescriptions of the Appendix meets the intent of the code. Even if local authorities have not adopted the Appendix, an application for an S-SIP wall system that clearly meets all of the prescriptions should not face any objections. However, if the limitations imposed by *Appendix S* do not accommodate a particular design, or if the local authorities have not adopted the Appendix, then alternative compliance will need to be sought.

Whether or not *Appendix S* is being used to justify a design, a particular S-SIP proposal may meet some code prescriptions but not others. In such cases, the permit application must clearly show the elements of the wall system that do conform to existing code prescriptions, and which elements require an alternative compliance pathway.

Alternative compliance applications

Every building code has a mechanism for consideration of non-conforming materials and approaches. If you have identified that some elements of your design cannot be supported via code prescriptions, it is incumbent on you to understand the exact procedures used in your code jurisdiction to handle alternative compliance. While the paperwork requirements will vary, all such alternative compliance pathways operate on the assumption that the applicant will provide proof that the alternative proposal meets or exceeds the provisions of the prescriptive code requirements. Any performance parameter (structural capacity, fire resistance, thermal performance, etc.) that exists for a wall in the prescriptive section of the code must be demonstrably met or exceeded by the proposed alternative. Each of these performance parameters must be fully supported and documented.

Several options exist for demonstrating that an alternative solution meets or exceeds code requirements:

Past performance An applicant can typically cite prior examples of the same or similar approach used successfully in the jurisdiction. Be sure to have adequate documentation of past performance to ensure that the approach was similar to what you are proposing, and to be sure that it was indeed a successful approach.

Testing data It is best if the tests are done to a code-recognized standard, such as ASTM, ANSI or CSA. If the tests are not performed to the standard used by the code, be prepared to show how the testing varies and how the results may be interpreted to show equivalency.

Professional seal A licensed architect and/or engineer can provide code equivalency assurance to the building department by applying their seal to the drawings and so ensure that to the best of their professional ability the alternative approach meets the intent of the code.

In some cases, it may be that all of these approaches are employed on an alternative compliance application. In all cases, understand that it is entirely up to you as the applicant to provide the information and any supporting interpretation to the building department. For better or for worse, building departments are reactive, not proactive. They are under no obligation to assist you with your documentation or ensure that it is complete. They are only obliged to respond to what has been provided in the application.

Code consultants are professionals that may be hired to assist an applicant with understanding the code and all the parameters that need to be addressed in order to put forward a complete application.

Rejections and appeals

It is important to know that a permit cannot be denied for any reasons other than code infractions or incomplete submissions. Every building code prescribes the manner in which a denial is presented to the applicant. In most code jurisdictions, the procedure for a permit refusal involves a written response explaining the code infractions that caused the permit to be denied. This is intended to give the applicant a full understanding of where the application was found to be lacking and provide a blueprint for resolving the issues in a resubmission. If all of the code issues are fully addressed in a subsequent submission, then a permit should be issued. In many cases, there can be several rounds of rejection and resubmission. While this may be frustrating and time consuming, getting a

building permit can be compared to taking a test where you must score 100%. Building departments cannot let any infractions they detect slip through without being addressed, so it is best to consider the application to be a multi-step affair. Forming and maintaining a good working relationship with the plan reviewer is very helpful. At best, the plan reviewer will be acting as an advocate and will be assisting you with understanding where the plans fall short of meeting the code and making suggestions regarding how the deficiencies can be corrected. At worst, they are obliged to make your mistakes known to you, and you will have to figure out how to correct them.

Should there be a disagreement about code compliance, every code jurisdiction has an established route for appeals. Often, this involves taking the dispute to the Chief Building Official. Should this fail to resolve the issue, there will be a higher regional, state or provincial authority that will hear appeals, and the pathway to accessing the appeal should be provided to you. Many appeal processes are quasi-judicial and involve a hearing where both the applicant and the building department put forth their arguments and a panel renders a decision. Here in Ontario, every case involving a straw bale wall system that has been appealed to the Building Code Commission has been ruled in favor of the applicant.

Preparation and patience are invaluable

Any application to a building department involving an alternative compliance element should be made well in advance of needing the permit to allow the process to go through a few rounds of back-and-forth. Expecting or, worse, demanding a fast turnaround for an alternative compliance application is to invite frustration and delays.

With thousands of straw bale projects already permitted, including numerous projects with S-SIPs, there is no reason for pessimism about receiving a permit. Any applicant willing to put the time and effort into making a complete initial submission and diligent enough to follow through with any requests for changes or more information should be rewarded with a permit.

Note:

1. *The Last Straw Journal*; California Straw Building Association (CASBA); Colorado Straw Bale Association (COSBA); Sustainable Sources Database; Ontario Natural Building Coalition; Straw Bale Association of Texas.

   Straw Bale Registry: www.sbregistry.sustainablesources.com/search.straw

Tools

S-SIPs CAN BE MANUFACTURED using typical construction tools. A panel building team will require:

- Measuring and marking tools.
- Compound miter saw or radial arm saw. A saw that can make cuts across the full width of framing members (typically 14 to 18 inches [355 to 460 mm]) will be valuable.
- Circular saw.
- Track saw or panel saw guide (will allow a typical circular to make accurate cuts for sheet goods).
- Impact driver.
- Table saw.
- Bale moisture meter.
- Grinder with lancelot wheel for cutting the width of bales.
- String line trimmer for smoothing the surface of bales.

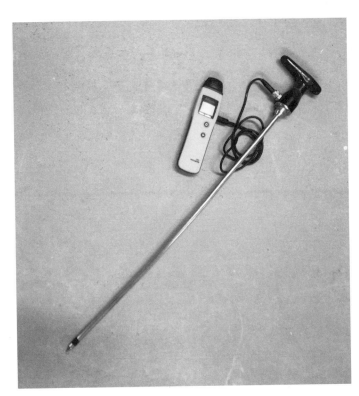

Moisture meter

String line trimmer.
CREDIT: KATIE HOWARD

For wet-process panels, plaster mixing and application tools will be needed:

1. Mortar mixer. Vertical shaft or horizontal shaft.
2. Wheelbarrow.
3. Floats and trowels.
4. Edger.
5. Pencil-style concrete vibrator

Vertical shaft mortar mixer.

Horizontal shaft mortar mixer.

A production facility would benefit from some specialty tools:

1. Panel saw. These allow fast and accurate cutting of sheet goods.
2. Plaster pump. For wet-process panels, a peristaltic pump can deliver plaster from the mixing station to the panels. This is helpful if the mixing is done outside and the panel production is indoors.
3. Forklift. Size the forklift for the type of panel weight and height. Fork extenders will allow the panels to be picked up lengthwise and widthwise.
4. Pallet jack. Allows panels to be moved manually without a forklift.
5. Portable saw mill. Allows multiple bales to be cut to height quickly and accurately.

Most portable band saw mills can be used to cut straw bales cleanly and accurately.

Chapter 15
Conclusion

PREFABRICATED STRAW BALE WALL PANELS are an idea in its early stages. The information presented in this book presents the direction and options that have been pursued by the early adopters of this approach. There are a great many improvements and developments yet to be made, and all of these will lower the labor input and material costs of S-SIPs. It is a simple idea, and will benefit from the enthusiasm of a larger pool of early adopters.

To date, S-SIP research and development of the approach has been a fraction of what is spent on other building products. What is remarkable is that S-SIPs, even in their infancy, rival products that have received decades of research and millions of dollars of investment. For a price that is already competitive with mainstream options, a handful of small S-SIP developers can already deliver a product offering great thermal properties, good carbon sequestering potential, excellent permeability and moisture handling capabilities, local and renewable materials, high indoor air quality potential, and low-tech, low-investment production.

I have often expressed the opinion that if the straw bale were patentable, the use of bales in construction would already be mainstream. Instead, bales are in the public domain, and so are the steps for creating successful S-SIPs. This is an idea that belongs to the commons, and it is an approach that suits adoption by the commons. The materials needed to make a viable S-SIP exist in most populated regions of North America, and the investment required to start making the panels is remarkably low.

For the individual owner-builder looking to make as ecologically friendly and energy-efficient a home as possible, while sticking to a reasonable budget and manageable timelines and labor input, it is difficult to find an approach that is as simple, approachable and do-able as building tilt-up S-SIPs on site.

The sun is just starting to rise on the potential for S-SIPs.
CREDIT: JANE STRONG

For the small-scale contractor or developer building a small number of homes or other buildings each year, the same advantages hold true. In addition, the ability to stage the wall construction off site and even off-season will allow for shorter timelines on the job site. S-SIPs can open up the market for affordable green building.

For an interested business owner or investor, a production facility for S-SIPs offers low entry costs and low training requirements to produce a high-value product. Even with current small-scale production, costs are very competitive. With additional mechanization and volume purchasing, it is feasible that S-SIPs could dramatically undercut the costs of other wall construction types, while offering true ecological benefits.

It has been exciting to be part of seeing this idea through its infancy over the past decade. With the publication of this book and the establishment of several full-time S-SIP production companies, the idea is entering a new phase. The more widely it is adopted, modified, examined, tested and re-tried, the stronger it will become and the better the likelihood of entering a more mainstream positioning.

There are not many examples in the building industry of an idea that offers measurable environmental benefits — including carbon neutrality or even carbon sequestration, chemical-free and waste-free production, nontoxicity, renewability, resilience and direct benefits to local economies — at a price that is competitive with more established but more ecologically damaging materials.

I encourage you to become part of seeing this idea into its adolescence and helping it to become a mainstream option by giving it a try and sharing your results and developments openly with your peers.

I look forward to the conversations...

Since watching my very first S-SIP get craned out of the barn, I've been excited by the potential of prefabricated straw bale walls.

Glossary

Terminology used in this publication uses the following definitions:

S-SIP. Straw bale Structural Insulated Panel; used to describe prefabricated straw bale wall panels.

BALE *or* **STRAW BALE.** A rectangular compressed block of *straw,* bound by *ties.**

DRY SYSTEM. A straw bale wall in which one or more of the *plaster skins* are replaced with a structural sheathing material. The sheathing provides a comparable degree of structural integrity, fire resistance, weather protection and *permeability* to a *plaster skin.*

LAID FLAT. The orientation of a *bale* with its largest faces horizontal, its longest dimension parallel with the wall plane, its *ties* concealed in the unfinished wall and its *straw* lengths oriented across the thickness of the wall.*

LOAD-BEARING WALL. A straw bale wall that supports more than 100 pounds per linear foot (1459 N/m) of vertical load in addition its own weight.*

MESH. An openwork fabric of linked strands of metal, plastic, or natural or synthetic fiber, embedded in plaster.*

NONSTRUCTURAL WALL. All walls other than *load-bearing walls* or *shear walls.**

ON-EDGE. The orientation of a *bale* with its largest faces vertical, its longest dimension parallel with the wall plane, its *ties* on the face of the wall, and its *straw* lengths oriented vertically.*

PERMEABILITY and **VAPOR PERMEABLE.** The property of having a moisture vapor permeance rating of 5 perms ($2.9 \times 10_{-10}$ kg/Pa • s • m^2) or greater, when tested in accordance with the desiccant method using Procedure A of ASTM E 96. A vapor permeable material permits the passage of moisture vapor.*

PIN. A vertical metal rod, wood dowel, or bamboo, driven into the center of stacked *bales,* or placed on opposite surfaces of stacked bales and through-tied.*

PLASTER. A wet-applied mixture of aggregate and binder (clay, lime or lime-cement) that hardens and is capable of providing structural, weather protection and fire resistant properties.

PLASTER SKIN. A continuous coating of plaster applied to a straw bale wall that is at least ¾ inch in thickness.

PLASTERED STRAW BALE. An assembly with a core of *straw bales* and interior and exterior *plaster skins* or *reinforced plaster* bonded to the straw bales that together provide an insulative and structural wall system.

PREFABRICATED. The construction of a wall such that structural and insulation elements are built as a unit prior to being placed upright and secured to the foundation.

REINFORCED PLASTER. A plaster containing *mesh* reinforcement.*

RUNNING BOND. The placement of *straw bales* such that the head joints in successive courses are offset at least one-quarter the bale length.*

SHEAR WALL. A straw bale wall designed and constructed to resist lateral seismic and wind forces parallel to the plane of the wall.*

SKIN. The compilation of plaster and reinforcing, if any, *wet-applied* to the surface of a straw bale wall.

STRUCTURAL WALL. A wall that meets the definition for a *load-bearing wall* or *shear wall.**

STACK BOND. The placement of *straw bales* such that head joints in successive courses are vertically aligned.*

STRAW. The dry stems of cereal grains after the seed heads have been removed.*

TIE. A synthetic fiber, natural fiber, or metal wire used to confine a straw bale.*

WET SYSTEM. A straw bale wall in which the inside and outside faces of the wall are coated with plaster to provide structural integrity, fire resistance, weather protection and *permeability.*

*Definitions from *Appendix S — Straw bale Construction, 2015 International Residential Code (IRC).*

Resources

S-SIP Producers

North America

Endeavour Centre

Peterborough, Ontario, Canada

endeavourcentre.org

Author Chris Magwood is a director at the Endeavour Centre, a sustainable building school. The school undertakes full-scale building projects, often using S-SIPs. They have also produced S-SIPs for clients and are currently undertaking a testing program for dry-process S-SIPs for commercial production. Endeavour offers hands-on workshops about S-SIP construction and consultation for owner-builders wishing to use S-SIPs.

mobEE

Guelph, Ontario, Canada

mobee.evolvebuilders.ca/

Evolve Builders Group offers prefabricated straw bale portable classrooms as a facet of their general construction services. These classrooms replace school portables made from unhealthy materials.

Europe

Ecococon

Vilnius, Lithuania

ecococon.lt/english

Ecococon designs and builds S-SIPs for residential and commercial projects, largely in Eastern Europe. They are currently working on regional production in other parts of Europe.

Ecofab

Cornwall, England

eco-fab.co.uk

Ecofab developed from site-built straw bale work undertaken by ARCO₂ Architecture. They moved to prefabrication strategies to meet the needs for low-income housing projects, and have continued to apply their technique to affordable housing and school projects.

Modcell

Bristol, England

modcell.com

Modcell is the longest established producer of S-SIPs, with numerous residential and commercial projects to their credit. They have also created a precedent-setting multi-unit residential building. They operate a permanent production facility, as well as "flying factories" set up close to build sites to minimize transportation.

Australia

SITUPS from Huff'n'Puff Straw Bale Constructions

Ganmain, NSW, Australia

glassford.com.au/main/sit-up-example/

Huff'n'Puff has a long track record of site-built straw bale homes, and began experimenting with their SITUP panels in 1998. They have built a number of S-SIP homes around Australia.

Codes

USA

Appendix S IRC 2015

In 2013, a ten-year effort to include straw bale building in the US building codes came to a positive conclusion when the International Code Council approved *Appendix S — Straw bale Construction* in the *2015 International Residential Code (IRC)*. This inclusion in the model code for the United States is a major step forward for the use of straw bale walls, including in prefabricated form.

A producer of S-SIPs can use this document to support code approval of a wall system that meets the requirements of the Appendix in the U.S. Where an S-SIP wall design does not conform to the Appendix, or in jurisdictions outside the U.S., alternative compliance pathways provided in local codes will need to be followed. Such alternative compliance applications may be supported using *Appendix S* and/or other relevant testing documents provided in the sections below.

The link to the full text of *Appendix S* is ecobuildnetwork.org/images/pdf_files/strawbale_code_support/AppendixS_SBConstruction_2015IRC.pdf.

Canada

ASRi (Alternative Solutions Resource Initiative) *The Straw Bale Alternative Solutions Resource.* (2013)

This publication contextualizes a wide range of research data into a format that is consistent with Part 9 of the National Building Code of Canada (residential construction). It follows the sections of the code and provides documentation that can be used in creating an Alternative Compliance application and is a highly recommended resource for S-SIP builders in Canada.

The Straw Bale Alternative Solutions Resource can be ordered in paper or PDF format from: asri.ca/shop.

Books

King, Bruce et al. *Design of Straw Bale Buildings.* Green Building Press, (2006).

Magwood, Chris et al. *More Straw Bale Building: A Complete Guide to Designing and Building with Straw.* New Society Publishers, (2005).

Magwood, Chris. *Making Better Buildings: A Comparative Guide to Sustainable Construction for Homeowners and Contractors.* New Society Publishers, (2014).

Racusin, Jacob Deva and Ace McArleton. *The Natural Building Companion: A Comprehensive Guide to Integrative Design and Construction.* Chelsea Green Publishers, (2012).

Journals

The Last Straw Journal

PO Box 1809, Paonia, CO, 81428, U.S.

thelaststraw.org

The Last Straw Journal is a quarterly publication dedicated to straw bale and natural building. Its back issues contain articles about S-SIPs and other issues relevant to panel builders.

Testing Data

A database of testing documents used to support the provisions in *Appendix S — Strawbale Construction* as approved by the International Code Council can be found at the California Straw Building Association website: www.strawbuilding.org.

The BuildWell Source is a free online library of alternative and natural building documentation. It has a comprehensive selection of straw bale wall tests and reports: buildwelllibrary.org/building-materials/fibers-1/straw-1/straw-bale-1.

Much of the testing found at BuildWell Source has been performed on straw bale wall units that closely resemble S-SIPs and is applicable to panelized construction. Some literature exists that is specific to prefabricated straw bale wall panels, generated at University of Bath in England:

Shea, A. D., Wall, K. and Walker, P. (2013) "Evaluation of the thermal performance of an innovative prefabricated natural plant fibre building system." *Building Services Engineering Research and Technology*, 34 (4). pp. 369–380.

Gross, Christopher. (2009) *Structural Performance of Prefabricated Straw Bale Panels.* University of Bath Department of Architecture and Civil Engineering.

Maskell, D. et al. (2014) "Structural development and testing of a prototype house using timber and straw bales." *ICE Proceedings Structures & Buildings*, 168 (1). pp. 67–75.

Wall, Katharine et al. (2012) "Development and testing of a prototype straw bale house." *Proceedings of the Institution of Civil Engineers: Construction Materials*, 165 (6). pp. 377–384.

Lawrence, M. et al. (2009) "Racking shear resistance of prefabricated straw-bale panels." *Proceedings of the Institute of Civil Engineers: Construction Materials*, 162 (3). pp. 133–138.

Appendix:
Engineering with Straw Bale Panels in Canada

WHEN STRUCTURAL ENGINEERS hear the term *straw bale panel*, what do we think of? We think of a straw bale SIP (structural insulated panel). Or using the acronym of this book, an S-SIP. A rectangular wall assembly consisting of a straw bale core inside a wooden frame with sheathing or plaster skin on each of the large faces. It is an assembly with much stiffer elements on its two faces than in its core. The wooden frame on the four edges can be customized to suit the installation.

To date, straw bale panels have more or less been limited to their use as wall assemblies (unlike "conventional" OSB-skinned, foam core SIPs, which are frequently used as floor and roof components as well as walls).

What does a structural wall need to do? The wall needs to support gravity (vertical) loads, and/or resist in-plane lateral loads. A structural wall (and exterior non-structural walls) must also resist out-of-plane lateral loads (wind and seismic). It may also need to resist uplift due to wind or in-plane rocking (for straw bale shear walls). All except gravity loads involve tension forces, and many of the load path geometries create eccentricity such that there is tension stress on one face of the panel (and compression on the opposite face). If there is a desire to implement an energy dissipating strategy for seismic design, rocking or sliding of individual panels or a group of panels may need to be accounted for. An engineer must also account for serviceability limits, such as maximum or minimum acceptable deflection of an assembly, including associated damage limits under short-term extreme loading (wind, projectile impact and seismic).

Engineering Design Options for S-SIPs

Detailed designs will not be provided here, nor will an exhaustive list of structural strategies for designing with S-SIPs. However, two strategies come to mind ahead of all others, and will be discussed here, with emphasis on the first method.

1. **Treating the straw bales as infill in a structural wood frame:** Since most bales are more than 24 inches (600 mm) long, the vertical members of most S-SIPs are spaced further apart than the maximum 24 inches (600 mm) allowed for conventional stud, "stick", or light framing in Part 9 of Canada's National Building Code (NBC) (and in Chapter 6 of the IRC in the U.S.). That said, it is possible for one side of each panel to include intermediate framing at no more than 24 inch spacing, and if structurally rated sheathing and connectors are used, then one side of the assemblies could be argued to conform with Part 9 of the NBC.

   If you cannot or do not want to put intermediate framing into the panels, the structural design will fall under Part 4 of the NBC, and is therefore required to conform to CSA O86, Engineering Design in Wood. (In the U.S. the structural design would fall under Chapter 23 of the IBC.) Whenever the spacing of vertical members is greater than 24 inches (600 mm), we move from prescriptive stud wall design to engineered post-and-beam design.

2. **Treating the straw bales and their plaster as a stressed-skin panel:** If enough is known

about the plaster skins, their bond to the bales, and the bales themselves, it is possible to conduct a strut-and-tie analysis on the composite system, treating each unit as a stressed-skin panel.

As the name suggests, the skins of these composite panels carry most of the stresses when the panel is loaded. The skins are generally much thinner but also much stiffer than the straw core, and therefore they "feel" the load first. The job of the core is to prevent the thin skins from buckling, and to connect and transfer loads between the skins.

Analyzing a plastered straw bale wall as a stressed-skin panel is not an easy task, since there is little data on the mechanical characteristics of straw bales or the connection between the surface(s) of the bales and the plaster skins with which to create a mathematical model that is representative of the final physical product. Then there is the question of the behavior of the connections between the panels and the respective floor, ceiling, and roof assemblies.

We are part of an international team working to produce enough information on bale and plastered bale behavior to develop a mathematical model based on fracture mechanics to describe panel behavior under various loads. Unfortunately, this effort is not well-funded, and is being carried out by well-meaning engineers and researchers separated by oceans, in their "spare" time. So for now, in terms of strict code compliant design in Canada, we recommend treating S-SIPs as a type of post and beam structure.

Which raises the following questions: What is the load path? Are both the interior and exterior framing designed as load-bearing elements? What about eccentricity? How do we distribute out-of-plane loading with posts spaced at 48 inches (1200 mm) or more?

Gravity loads and out-of-plane loads

The interior, exterior, or both sides of a panel can be designed and constructed to be structural. There may be reasons for selecting any one of these options. It may be advantageous to use a ledger and/or hanger connection for a floor system that bears on the interior side of the panels. Where a truss roof system has a large overhang, it may make sense for the exterior to be the bearing side. For heavy axial loads (centered on the wall) where the bearing can be considered uniform across the width of the panel, using both interior and exterior framing and/or skins is sensible. This is to account for out-of-plane loading as a combined axial + bending stress on the built-up posts at the connections between panels, and to reduce potential eccentricity. This requires anchorage at eight points on each panel — at all four corners on the top and the bottom. It can produce a tension load on one face and a buckling load on the other, and care should be exercised when using this strategy.

With panels starting at 16 inches (400 mm) thick, out-of-plane load capacity is generally not an issue because of the panels depth, with connection capacity governing in almost all cases.

In-plane loads

Shear walls

CSA 086 has design guidelines for sheathed stud walls, and there are a variety of shear walls defined as shear force resisting systems (SFRS) with allowable ductility and over-strength coefficients in the NBC, including stud walls sheathed with gypsum panels (although only in combination with wood-based panels — i.e., exterior wood-based panels and interior gypsum panels).

The ductility and over-strength coefficients assigned to these wood shear walls are based on nailed connections and the relatively well-known yield and deformation behavior of nailed wood connections under cyclic shear loading.

King et al. have suggested that tests on a limited number of plastered straw bale panels show a similar enough pattern of energy absorption and limited deflection/distortion under cyclic loads to use the ductility and over-strength coefficients for a nailed shear wall given in the 1997 Uniform Building Code (UBC) when designing a plastered straw bale panel. King suggests that this is conservative based on observation of panel behavior under extreme loading in test conditions, and also of observed "tough-ness" of plastered straw bale walls in service.

We have been challenged by colleagues in the international engineering community on the appropriateness of this design strategy, and have been supported by others. We cannot decide for each practitioner, but can say that using a combined ductility and over-strength coefficient of 1.3 and designing from first principles is allowed under the NBC, and is definitely conservative. Broadly speaking, this keeps the structure elastic at all times — which is well and good — but in addition plastered bale wall assemblies have shown tremendous capacity to absorb energy at deflections past their elastic limit without catastrophic failure in shake table, monotonic, and reverse-in-plane-cyclic tests. We leave it to individual engineers to decide how well they can translate this information into relevant design data for their particular project.

Braced frames

CSA 086 does not provide detailed guidance on the design of braced frames. The NBC does allow wooden braced frames for seismic force resisting systems, and assigns ductility and over-strength coefficients based on the ductility of the connections used in the frame. CSA 086 does allow design with a minimum combined ductility and over-strength coefficient of 1.3 for wood frame structures not explicitly defined in the standard when the ductility of connections cannot be demonstrated satisfactorily.

Anchorage for the braced frame design needs to positively connect vertical members to the floor/foundation below and the floor/roof above each panel, and may need to be a relatively robust tie-down for seismic or high wind loading situations, as well as an adequate connection for transfer of shear loads to the floor and roof diaphragms.

Components:

1. Base plate

These are generally 2× dimensional lumber and can either be laid flat (most common) or on edge. Sill plates for conventional stud walls are generally 2× material laid flat and connected to a foundation below using anchor bolts at a specified spacing, usually less than 48 inches (1200 mm). When the sill plate is connected to a floor system above a foundation, nail spacing is specified, and further anchorage may be in place as required. One or both of the base plates on the S-SIP can be placed so they are adjacent to the sill plate and structural screws can be used through the base plates to connect to the sill plate. If the structural design requires tie-downs, these can be incorporated as part of the side plate-to-base plate connection.

BASE PLATES

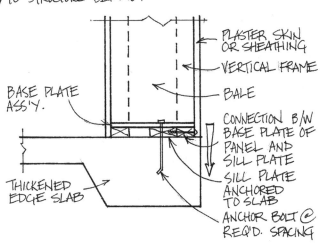

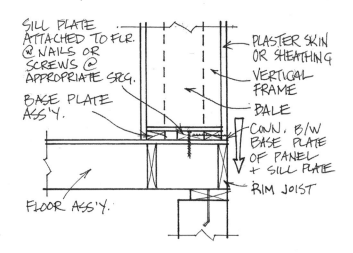

(A) SILL PLATE ON SLAB

(B) WOOD FLR. ASS'Y.

In cases where there are significant seismic loads, the plaster mesh may require more than 1½ inches (40 mm) of plate thickness to give the connection the appropriate capacity. In these cases, King suggests using a 4×4 base plate (based on cross-grain failure of nominal 2× plates during in-plane lateral load testing). There is no reason that a deeper base plate cannot be connected to a deeper sill plate in the case of an S-SIP assembly.

The connection between the base plate assembly of the S-SIP and the floor/foundation must be sized to transfer the design loads from the panels into the structure below. This is very simple if there is no net up-lift load and the building is in a low-seismic zone (Seismic Design Categories A and B in the IBC and IRC). However, in cases where there is moderate or high seismic loading (Seismic Design Categories C and D in the IBC and IRC) or significant wind loading (greater than 140 mph ultimate design wind speed) or the floor is being used as a diaphragm, all of these forces must be accounted for.

It is quite possible that the axial capacity of a given design will be limited by bearing area due to the compressive resistance perpendicular to the grain of the base plate. In our experience, this capacity is not significantly lower than the calculated capacity of a two-ply built-up post made from the same material. However, it may be worth using higher density material in the base plate to utilize the full design axial capacity of the side plates.

2. Side plates

An efficient way to use the adjacent side plates between panels under axial loading is to connect them to create a built-up post. The challenge posed by prefabricated panels is to achieve a conforming connection when the face of the

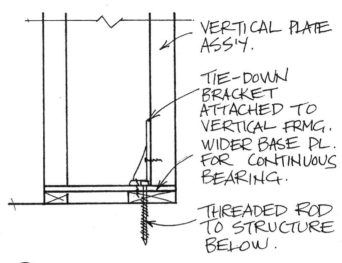

© TIE-DOWNS

VERTICAL PLATE ASS'Y.

TIE-DOWN BRACKET ATTACHED TO VERTICAL FRMG. WIDER BASE PL. FOR CONTINUOUS BEARING.

THREADED ROD TO STRUCTURE BELOW.

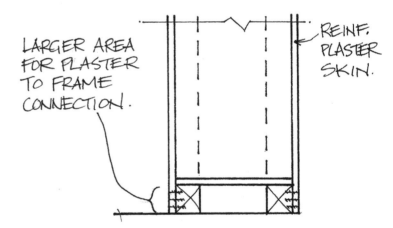

Ⓓ 4×4 BASE PLATES

LARGER AREA FOR PLASTER TO FRAME CONNECTION.

REINF. PLASTER SKIN.

adjacent piece of lumber is not accessible. CSA 086 does allow small diameter connectors to be installed at an angle (so-called "toe-nailing," or "toe-screwing"), but at reduced capacities. It may be difficult to guarantee that minimum edge distances are achieved using connectors at an angle. At the same time, many European structural screw manufacturers encourage installation of their product at an angle to ensure the fastener is in tension, reducing the risk of a brittle failure.

The use of sheathing in built-up side plates greatly assists the out-of-plane bending strength of each panel, as either a box-beam or I-beam section is created. Appropriate connector spacing within each beam element and attachments at the top and bottom of each pair of panels to handle shear transfer are straightforward.

Prefabricated I-joists are not typically rated for axial loading, and often use finger-jointed material in the flanges. An assembly using two I-joists side-by-side could be designed as a pair of built-up posts, with appropriate strength characteristics assigned based on the grade of material and the connectors used. I-joists do have excellent bending characteristics, and are very dimensionally stable. If the flanges are made from graded dimensional lumber, then the axial capacity can be determined from first

principles, taking into account the reduced cross-section from the web.

3. Top plate

The simplest top plate is a mirror image of the base plate: two 2× dimensional lumber members running along the edge of the panel. Where a 3× or 4× base plate is used, a 2× top plate may be sufficient, as there is often less demand at the top of the wall.

It is often beneficial to construct the top plate as a type of box beam in order to transfer load from joists, rafters or trusses that are spaced tighter than the width of the panel into the side plates. Alternatively, a separate beam structure can be used to connect adjacent panels and to span above window or door openings, if the openings are not within panels themselves.

In panels where there is a built-in window or door opening, the top plate must either serve

BOX BEAM SPANNING OPENING BETWEEN PANELS.

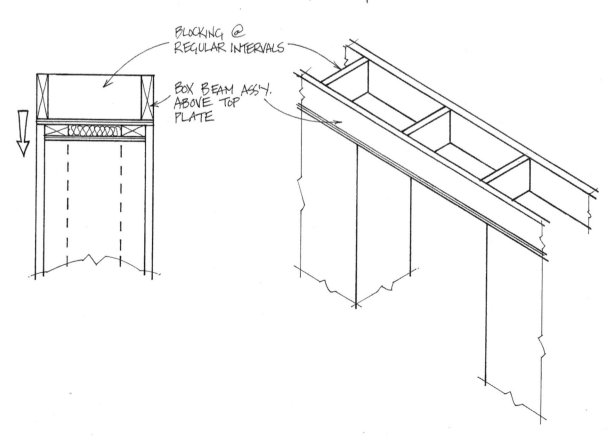

BLOCKING @ REGULAR INTERVALS

BOX BEAM ASS'Y. ABOVE TOP PLATE

as a lintel, or transfer its load from a lintel to the rest of the structure.

If the building is in a moderate or high seismic zone, the connection between the top of the panels and the ceiling/roof diaphragm must be sufficient to transfer the shear load in both principle axes.

4. Straw bales

It is obvious to anyone who has worked with straw bale wall systems that the presence of the bales in the assembly does more than simply provide insulation value. A plastered bale wall is definitely more than the sum of its parts. But how much more? Is it reasonable to consider vertical framing members in panels as continuously laterally braced on their weak axis by the plaster/bale combination? When building "dry" panels, is the bracing effect of the straw bale infill on the vertical framing significant, compared with the fastener/sheathing connection to that framing?

There are two compelling reasons to justify NOT using the additional capacity of the bales in the structural design of an infill wall panel (aside from some lateral bracing action for the vertical framing members):

1. Load flows to stiffness — In all cases, the framing, sheathing, plaster and combination of these elements is stiffer than the bales. Loads will flow through a path along these stiffer elements, and not through the bales — at least not until extreme plastic deformation occurs.

2. Creep — At densities that keep bales in the human scale for handling (below 15 pounds per cubic foot), baled straw exhibits more creep behavior than dimensional lumber and (most) plasters or sheathing products. This means that any long-term load that the bale actually does "feel" at the time of construction will cause permanent deformation of those bales. If the panel is well designed, that deformation will create a new, safe load path between the other elements in the assembly.

5. Sheathing

Manufacturers provide design data for sheathing products, but most of this data is based on testing a stud wall assembly — with regularly spaced framing at 24 inches (600 mm) or less. This may leave a designer grasping for defensible values to use for lateral load capacity.

In cases where sheathing cannot be counted on for lateral load capacity, internal diagonal bracing can be incorporated. CSA 086 allows for steel elements or dimensional lumber to be used as a tension-resisting part of a braced frame assembly.

(*Note*: We are not providing specific guidance on the structural design of panels using the strength of plaster skins, for reasons stated above with respect to stressed-skin panels. We refer the reader to Bruce King's *Design of Straw Bale Buildings*.)

6. Connectors

Just as important as the sheathing or plaster used for the skins of the panel is the connection of those skins to the frame and/or bales. In Canada, there are many wood screws and nails manufactured offshore to no particular standard. These may or may not be suitable for structural use, and thus are difficult to use in a structural design with certainty. That said, there are standards for nails, screws and staples that are referenced in the NBC, in particular ASTM F-1667. Connectors manufactured and tested to this standard will have technical information available for structural design.

Summary

Since the 2005 code cycle, the NBC has allowed for alternative solutions, and all of the provincial and territorial codes adopting variations of the national model code have retained that provision in one form or another. Municipalities and authorities having jurisdiction have protocol in place for applications including alternative solutions, or at the very least, they are supposed to. It is possible for a qualified designer to carry out a rational structural analysis and design outside of the CSA 086 design standard and for the design to be reviewed, permitted, built and inspected.

However, for designers and engineers looking for guidance on relatively straightforward code compliant structural design of straw bale panels (S-SIPs) in Canada, the information presented here will give them a place to start.

— Tim Krahn, MSc, P.Eng. and
Kris Dick, Phd, P.Eng., Building Alternatives Inc.,
January 2016

Index

About the Author

CHRIS MAGWOOD is obsessed with making the best, most energy efficient, beautiful and inspiring buildings without wrecking the whole darn planet in the attempt.

Chris is currently the executive director of The Endeavour Centre, a not-for-profit sustainable building school in Peterborough, Ontario. The school runs three full-time, certificate programs: Sustainable New Construction, Sustainable Renovations and Sustainable Design, and it hosts many hands-on workshops annually.

Chris has authored numerous books on sustainable building, including *Making Better Buildings* (2014), *More Straw Bale Building* (2005) and *Straw Bale Details* (2003). He is co-editor of the Sustainable Building Essentials series, and is a past editor of *The Last Straw Journal,* an international quarterly of straw bale and natural building. He has contributed articles to numerous publications on topics related to sustainable building and maintains a blog entitled "Thoughts on Building."

In 1998 he co-founded Camel's Back Construction, and over eight years helped to design and/or build more than 30 homes and commercial buildings, mostly with straw bales and often with renewable energy systems.

Chris is an active speaker and workshop instructor in Canada and internationally.

If you have enjoyed *Essential Prefab Straw Bale Construction*, you might also enjoy other

BOOKS TO BUILD A NEW SOCIETY

Our books provide positive solutions for people who want to
make a difference. We specialize in:

**Food & Gardening • Resilience • Sustainable Building
Climate Change • Energy • Health & Wellness • Sustainable Living**

**Environment & Economy • Progressive Leadership • Community
Educational & Parenting Resources**

New Society Publishers

ENVIRONMENTAL BENEFITS STATEMENT

New Society Publishers has chosen to produce this book on recycled paper made with
100% post consumer waste, processed chlorine free, and old growth free.

For every 5,000 books printed, New Society saves the following resources:[1]

36	Trees
3,243	Pounds of Solid Waste
3,568	Gallons of Water
4,654	Kilowatt Hours of Electricity
5,895	Pounds of Greenhouse Gases
25	Pounds of HAPs, VOCs, and AOX Combined
9	Cubic Yards of Landfill Space

[1]Environmental benefits are calculated based on research done by the Environmental Defense Fund and
other members of the Paper Task Force who study the environmental impacts of the paper industry.

For a full list of NSP's titles, please call 1-800-567-6772 *or check out our website* at:

www.newsociety.com